DATE DUE			
Nov 1 7 8			
Mar27 '80			
Apr 15 '81			
Mar29 '82			
Apr 30 '82			

The Economics of Nuclear and Coal Power

Saunders Miller
assisted by
Craig Severance

The Praeger Special Studies program—utilizing the most modern and efficient book production techniques and a selective worldwide distribution network—makes available to the academic, government, and business communities significant, timely research in U.S. and international economic, social, and political development.

The Economics of Nuclear and Coal Power

PRAEGER SPECIAL STUDIES IN U.S. ECONOMIC, SOCIAL, AND POLITICAL ISSUES

Praeger Publishers   New York   Washington   London

Library of Congress Cataloging in Publication Data

Miller, Saunders.
    The economics of nuclear and coal power.

    (Praeger special studies in U.S. economic, social,
and political issues)
    Bibliography: p. 147
    1. Atomic power industry—Costs.   2. Electric power-
plants—Costs.   3. Electric power-plants—Fuel consumption.
4. Coal—Costs.   I. Title.
HD9698.A2M535   338.4'3   76-24361
ISBN 0-275-23710-9

PRAEGER PUBLISHERS
111 Fourth Avenue, New York, N.Y. 10003, U.S.A.

Published in the United States of America in 1976
by Praeger Publishers, Inc.

Printed in the United States of America

The choice of energy sources for new large scale electric generating plants presently has narrowed to an option of either coal or nuclear power. There is a sharp divergence of opinion regarding this option, not only between environmentalists and industry, but within the utility industry itself. Many utilities view nuclear energy as the utimate answer to safe, cheap electricity while opponents foresee disaster because of an alleged lack of safety both in the plants themselves and in the handling of radioactive fuel and waste. Yet, a large body of environmentalists stand opposed to coal energy, viewing it as a blight to the landscape and as harmful to the health of man because of air pollution. In both situations, the arguments are highly charged, and all protagonists deal in qualitative verbiage and technical argot, armed with esoteric and frequently questionable statistics.

As the authors examined publication after publication on the subject of nuclear and coal energy, it became most apparent that the economics of both energy sources were ignored or glossed over. Yet, in the final analysis, except for certain regulated industries, government subsidized programs, and other politically supported projects (such as obsolete weapons systems and pork barrel public works), all business decisions in the long run within a capitalist society are (or should be) determined by the economics of profit and loss, which in turn are governed by the laws of supply and demand.

It seemed logical, therefore, to conclude that the answer to the coal vs. nuclear controversy would probably lie within the framework of economic analysis. At the inception of the study, it was anticipated that the assemblage of the data would be a relatively mechanical task and that a few computations with current data would lead to a simple comparison.

This, however, was not the case. The material on nuclear energy was found to be almost the exclusive province of the Energy Research and Development Agency (ERDA) and its predecessor, the Atomic Energy Commission (AEC). And, while all the information was generally available, there were numerous problems in utilizing it:

1. It was scattered in bits and pieces throughout numerous publications;
2. The information was never tied together in such a manner that firm conclusions could be drawn;
3. In working from raw data to economic conclusions, there were large gaps in the methodology that led to the final problem;
4. Although the raw data was accurate, the assumptions used in generating derived data, especially regarding the economics of nuclear power, were often weak, glibly optimistic, or outdated.

It was observed that most articles published on the economics of nuclear energy, whether in general business magazines or in trade publications, relied upon historical data that currently bears no relation to the cost of projects begun at this time. As for coal power, it was found that the economic aspects were almost ignored, much as a stepchild who received scant attention.

The authors of this report have sought to remedy the historically inadequate economic analysis performed by both utilities and government, and have sought to explore on a rigorous financial basis the comparative costs of these two major power sources with attention to their effect on the future of electric utility companies. The authors have also introduced the concept of "economic risk" into the analysis.

However, in evaluating any investment, it is not sufficient merely to select the option that has the lowest cost. The concept of "economic risk" must also be considered. This is the possible loss of some degree of income that occurs because an investment does not perform as planned.

A simplified example may serve to convey the concept. Assume that a traveler must go from Los Angeles to New York by car. He owns no car at the beginning of the trip and has no need for a car after completing the journey. His first option is to buy an old car for $200 and scrap it in New York. The second option is to buy a car for $3,000 with a guaranteed sale price of $2,700 at the other end. Thus, the cost would be $200 with the "clunker" and 50 percent more, or $300, with the newer car. But would the "clunker" achieve its purpose? Would it get the traveler to his destination, safely, if at all? And, especially, what would be the economic loss from delays encountered on the journey? If these factors are considered in the decision-making process, then the choice becomes obvious. "Cheapest" may not really be cheapest at all.

In much the same way, the loss from "downtime" (i.e., nonoperating time) and the loss from any failure to perform according to plan must be included in the selection procedure for a power generating plant. All of those elements that could reduce the capability of a plant to deliver the design level of power must be an integral part of the cost analysis.

Furthermore, the utility executive must also include in his analysis the effects of the "maximum probable adverse" situation that could occur. This would exclude, perhaps, the effect of a meteor striking a plant because this probability may be too infinitesimal to include in an analysis. However, the probability of being without fuel may be a relevant one, and this must be included in the final determination of what type of plant to build. If the probabilities of negative occurrences and the costs associated with these occurrences are not quantitatively factored into the final costs, then improper and fallacious decision making has occurred.

In the analysis that follows, an attempt was made to be as conservative as possible at every step. As a result of this conservatism, any technology that was either untried commercially or not yet in use in the United States was excluded from consideration. This would include such activities as the Batelle coal desulfurization process, magnetohydrodynamics, and long-wall mining, an underground method used in Europe with recovery rates as high as 85 percent, as compared to the 50 percent rate assumed in this report. Neither were underground or offshore nuclear reactors or fusion power assumed to be available. Also, high-temperature gas-cooled reactors (HTGRs) were not considered due to the fact that, except for one prototype, it seems unlikely that more will be built.* For coal-fired plants, the economies to be achieved via "mine-mouth" plants were not factored in, nor were the additional revenues realizable by locating a coal plant adjacent to a city and selling the by-product steam taken into account.

It should be understood that this book relates to decisions regarding plants that will come on-stream in six to ten years. Thus, when analyzing alternative investments, the problems of pollution and sulfur emissions from coal, for example, must be viewed within the context of whether or not the expectation is that these problems will be resolved by the mid-1980s.

As will be seen, this study deals with elements of a rapidly changing economic environment. Changes in costs and in the availability of materials and services are taking place daily. Most of the reference data in this book was gathered in 1975; and given the long lead times of book publishing, the question arose whether to attempt an update of the data just prior to publication if it meant using less authoritative sources which could not be validated. Bearing in mind that a major goal of this book is to provide a reference document which can be used as the basis for performing an accurate economic analysis for any geographic region, factoring in with ease new events as they occur, and that by mid-1976 no significant changes had occurred in the overall economic parameters, the decision was made to retain all of the original reference material and merely to footnote some changes.

The methodology used has been explicitly presented in a step-by-step form in order that, as exogenous conditions change, updates can easily be performed by concerned parties. We hope this book will succeed in putting the question of power economics into an accurate perspective that will result in sound investment decisions and in the establishment of realistic government policies and priorities.

---

*According to the *Wall Street Journal* of February 25, 1976, Gulf Oil Corp. "has lost over $600 million before tax benefits" on the Fort St. Vrain, Colorado HTGR, and its partner Royal Dutch Shell Group "has set aside over $300 million for losses."

CONTENTS

                                                                    Page

PREFACE   v
LIST OF TABLES   xi
LIST OF FIGURES   xiv

Chapter

  1   URANIUM DEMAND AND SUPPLY   1

        Uranium Demand   1
        Uranium Supply   5
        Comparison of Uranium Supply with Demand   8
        How Probable is Significant Uranium Expansion?   12
        Foreign Resources   17
        Summary   20
        Conclusion   20
        Notes   21

  2   NUCLEAR FUEL CYCLE   22

        Mining and Milling   23
        Conversion to UF_6 30
        Enrichment   30
        Nuclear Fuel Processing and Fabrication   37
        Reprocessing of Spent Fuel   38
        The Problems of Spent Fuel Storage   43
        Radioactive Waste Mangement   44
        Summary   45
        Conclusion   46
        Notes   47

  3   COAL AVAILABILITY AND COST   48

        Adequacy of Coal Reserves   48
        Barriers to Coal Usage   50
        Coal Demand   51
        Demand vs. Supply   52
        Coal Prices   54

Chapter		Page
	Availability of Capital and Equipment	56
	The Transport of Coal	56
	Summary	59
	Conclusion	59
	Notes	59
4	HOW THE PRICE OF ELECTRICAL POWER IS DETERMINED	61
	The Theory of Rate Determination	61
	Generation Cost vs. Delivered Price	63
	Factors Comprising Generation Costs	64
	Conclusion	66
5	CAPACITY FACTORS	67
	Definition of Terms	67
	The Effect of Capacity Factors on Electricity Costs	68
	Historical Results	68
	Interpreting the Data	71
	Summary	75
	Conclusion	75
	Notes	76
6	ANALYSIS OF BASE YEAR COSTS	77
	Uranium Cost	77
	Conversion to UF_6	78
	Enrichment	78
	Reconversion and Fabrication	80
	Spent Fuel Shipping	80
	Reprocessing	81
	Waste Management	82
	Coal Fuel Costs	83
	Operating and Maintenance Costs	83
	Construction Costs of Nuclear and Coal Plants	83
	The Effects of Capacity Factors	84
	Summary	87
	Conclusion	87
	Notes	88
7	PROJECTING FUTURE COSTS	89
8	THE BREEDER REACTOR	93

Chapter   Page

   9   CAN UTILITIES SURVIVE MASSIVE CONSTRUCTION
       PROGRAMS?   97

          Seeing the Forest, Not the Trees   97
          The Vicious Cycle   98
          Short-Term and Intermediate-Term Risks   99
          The Viewpoint of a Regulatory Agency   100
          The Potential for a Bailout   101
          Mirrors and Mickey Mouse Solutions   101
          Conclusion   101
          Notes   102

  10   CONCLUSION   103

          Note   109

Appendix

   A   FORECASTS OF ENERGY CONSUMPTION & ELECTRIC
       GENERATING CAPACITY   110

   B   THE MISSING LINKS   124

   C   INTERVIEWS REGARDING COAL   125

   D   CALCULATING OVERNIGHT PLANT COSTS AND
       CONSTRUCTION FACTORS   127

   E   METHODOLOGY FOR COMPUTING CAPACITY FACTORS   129

   F   SOLAR POWER   131

   G   WASTE AS FUEL   142

BIBLIOGRAPHY   147

ABOUT THE AUTHORS   151

LIST OF TABLES

Table		Page
1.1	Annual Growth Rates In U.S. Energy Consumption	1
1.2	Total Energy Projections	3
1.3	Electricity Growth	4
1.4	Installed Nuclear Electrical Generating Capacity	4
1.5	Uranium Requirements in Metric Tons of U_3O_8	6
1.6	Metric Tons U_3O_8	9
1.7	Reactors Supportable by Uranium Resources	11
1.8	Number of 1,000-Megawatt Plants Beyond Carrying Capacity of Fuel Supply Projected to Begin Construction	12
1.9	Indicators of Trend of Future U.S. Uranium Resource Development	19
2.1	Ore Processing Capacity of Uranium Milling Companies and Plants	24
2.2	Exploration/Mining/Milling Factors Associated with Meeting Contracted Requirements for the Operating Plan	28
2.3	Exploration/Mining/Milling Factors Associated with Production to Fill Contracted Requirements	29
2.4	Separative Work at 0.30 Tails	32
2.5	Enrichment Capacity and Capital Requirements at 0.30 Tails Assay	38
2.6	Reprocessing Capacity and Capital Requirements	42
3.1	Total Coal Resources	49
3.2	Annual Coal Demand for Electrical Generation	51
3.3	Maximum Annual Coal Demand Assuming No Nuclear Expansion After 1985	52

Table		Page
3.4	Cumulative Coal Demand for Electrical Generation Based on ERDA Projections	53
3.5	Year of Coal Resource Depletion (50 Percent Recovery Rate)	53
3.6	Typical Coal Cost	55
3.7	Coal Transportation in 1972	56
4.1	Example of Cost of Capital Computation	62
5.1	Hypothetical Example of the Effects of Capacity Factors Upon Capital Costs	69
5.2	1973-74 Capacity Factors by Age of Plant for Boiling Water and Pressurized Water Nuclear Reactors	69
5.3	1961-73 Coal-Fired Plant Capacity Factors vs. Age of Plant	70
5.4	Capacity Factors by Size of Plant	71
6.1	Hypothetical Illustration of the Use of Probability Analysis in Computing Capital Costs	86
7.1	Construction Factors	91
7.2	Selected Capital Recovery Factors	92
8.1	LMFBR Power Capacity in Megawatts	95
A.1	Forecast of U.S. Energy Consumption and Electric Generating Capacity	110
A.2	Electrical Generating Capacity, Generation, and Energy Consumption	112
A.3	Forecast of Energy and Electricity in the United States	117
A.4	Fuel Cycle Lead Times	118
C.1	Results of Survey of Coal Operators (March 1975)	126
D.1	Survey of Nuclear Plant Extimated Completion Costs	127

Table   Page

D.2   Calculation Factors & Midpoint of Funds Expenditure for a
        Nuclear Plant   128

F.1   Incremental Nuclear Generation, 1985-2000   138

F.2   Square Miles Required for Solar Cell Systems in Year 2000   139

G.1   Cost Comparison of Waste as Fuel with Coal   145

LIST OF FIGURES

Figure		Page
1.1	Diagram of Energy Flows	2
1.2	Exploration Activity—Reserves Additions	14
1.3	U.S. Annual Mineral Production—Growth Comparisons	15
1.4	Depletion of $15 Reserves as $8 Reserves Are Mined	16
1.5	Non-Communist Annual Foreign Requirements and Production Capability	18
2.1	Estimated Schedule of Maximum Reasonable Production Capability from $8 Resources Compared with U.S. Separative Work Commitments	25
2.2	Projected Maximum Reasonable ('Could Do') U_3O_8 Production Capability of the Uranium Mining Milling Industry Through 1985	26
2.3	$8 Production Capability Scheduled to Meet Domestic U_3O_8 Requirements	27
2.4	Separative Work Example	32
2.5	Annual Separative Work Production Capabilities Versus Contracted Commitments	33
2.6	3-Site Average Annual Power Levels	34
2.7	Typical Lead Times in Enriched Uranium Supply	36
2.8	LWR Spent Fuel Reprocessing Supply and Demand (Cumulative Basis)	40
2.9	Comparison of Industry Reprocessing vs. Domestic Uranium Returns	40
2.10	Industry Processing Plans	41
5.1	Coal and Nuclear Capacity Factors by Age of Plant	72

Figure		Page
5.2	Coal and Nuclear Capacity Factors by Size of Plant	73
A.1	Normal Uranium Feed Requirement	121
A.2	Separative Work Requirement	122
F.1	Solar Heated House	132

The Economics of Nuclear and Coal Power

1

URANIUM DEMAND
AND SUPPLY

The starting point for any business venture is to be assured that the raw material required for production is available in sufficient quantity and at an economically attractive price. This same logic applies to the electric power industry.

URANIUM DEMAND

ERDA has projected future Gross National Product (GNP) based on econometric analysis, which takes into account population, employment, and productivity factors. Energy consumption is then related to this model, using the energy network system developed by the Brookhaven National Laboratory shown in Figure 1.1.[1] Table 1.1 then shows the growth rates that were computed as a function of the scenario development. This translates into the energy requirements given in Table 1.2.

TABLE 1.1

Annual Growth Rates in U.S. Energy Consumption
(in percent)

	High	Moderate	Low
1975-85	3.7	2.75	2.0
1986-2000	3.5	3.45	2.3

FIGURE 1.1

Diagram of Energy Flows

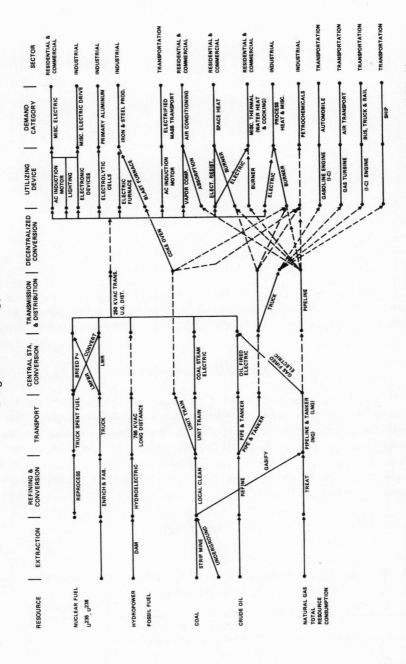

Source: U.S. Atomic Energy Commission, *Nuclear Power Growth 1974-2000*, WASH 1139 (74), (February 1974): 10.

2

TABLE 1.2

Total Energy Projections (10^{15} Btu)

	1973	1980	1985	1990	2000
High	75.6	95.3	116.6	136.8	195.0
Moderate	75.6	89.7	104.8	122.6	174.3
Low	75.6	86.1	96.2	107.9	135.3

Source: Energy Research and Development Administration, Office of the Administrator for Planning and Analysis, *Total Energy, Electric Energy, and Nuclear Power Projections*, February 1975.

As can be seen, three energy patterns were developed. The first of these is the "high case," based on continuing past historic growth rates with no emphasis on energy conservation. The "moderate case" assumes some conservation measures together with improved efficiencies in energy use. In the "low case" a slower rate of economic growth is envisioned. Also, energy prices are assumed to rise more rapidly than those of other commodities; there is decreased emphasis on the production of goods, and increased efforts are made to achieve energy conservation.

From this overall forecast of energy demand, a projection of electricity growth was derived as shown in Table 1.3. The following assumptions were made:

Electric energy represented 26 percent of total energy inputs in 1973. High case: The 7 percent historic growth rate in electricity demand resumes through the mid-1980s and then falls off to 6.4 percent in the year 2000. By then, electric energy inputs increase to 50 percent of total resource consumption.

Moderate/high case. Electric production grows at a 6.25 percent rate through 1985, and then falls to a 5.85 percent rate. Rising electric prices are expected to cause reductions in future demand, but because of the reduced availability and the relative prices of other fuels, substitution of electricity takes place.

Moderate/low case. Substitution occurs at a more modest rate because electric prices are not so advantageous and other fuels are more readily available than in the moderate/high case. Electric production grows at a 6.0 percent annual rate through 1985 and then slows to 5.4 percent.

Low case. Stringent conservation methods are imposed, but electricity represents a higher percentage of energy demand. Kilowatt-hour growth is 5.8 percent through 1985 and 4.75 percent thereafter, but electric energy inputs account for 50 percent of total energy inputs by the year 2000.

TABLE 1.3

Electricity Growth (Billion Kilowatt-Hours)*

	1973	1980	1985	1990	2000
High	1,878	2,780	3,905	5,290	9,880
Moderate/High	1,878	2,675	3,660	4,820	8.600
Moderate/Low	1,878	2,630	3,570	4,660	7,925
Low	1,878	2,570	3,500	4,400	7,020

Source: U.S. Energy Research and Development Administration, Office of Administrator for Planning and Analysis, *Total Energy, Electric Energy, and Nuclear Power Projections*, February 1975.

*Kwh (kilowatt-hour) is 1,000 watts of electricity for one hour of time.

Based on these figures, ERDA then computed a forecast of energy consumption and electric generating capacity by type of power unit; for example, nuclear, fossil, hydroelectric, and so on. (The complete projections are given in Appendix A.)

TABLE 1.4

Installed Nuclear Electrical Generating Capacity (GWe*)

	Low Case	Moderate/Low	Moderate/High	High Case
1973	18.4	18.4	18.4	18.4
1980	70.5	76.0	82.0	92.0
1985	160.0	185.0	205.0	245.0
1990	285.0	340.0	385.0	470.0
1995	445.0	545.0	640.0	790.0
2000	625.0	800.0	1,000.0	1,250.0

Source: U.S. Energy Research and Development Administration. Office of the Administrator for Planning and Analysis, *Total Energy, Electric Energy, and Nuclear Power Projections*, February 1975.

*GWe is "gigawatts" of electricity. A gigawatt is 1000 megawatts or 10^6 kilowatts. According to "Uranium Reserves, Resources, and Production" dated June 15, 1976 1976 published by the Federal Energy Resources Council citing "National Energy Outlook," Federal Energy Administration, FEA-N-75/713, February 1976, expected nuclear capacity will be 80 GWe for 1980, 150 GWe for 1985 and 225 to 300 GWe for 1990. As can be seen, these projections lie within a range similar to those developed by ERDA in 1975.

From this data, ERDA then forecast uranium requirements for each of the four growth levels under two different operating environments—without plutonium recycle, and with plutonium recycle. Except for the "high case," where the forecast was available solely under plutonium recycle conditions, only a "no plutonium recycle" environment has been used.* A list of the annual and cumulative uranium requirements is presented in Table 1.5.

With potential demand forecast, it now becomes relevant to determine how much of this demand can actually be met by known supplies.

URANIUM SUPPLY

The nuclear fuel cycle begins with the mining and milling of uranium ore; a key concern in the entire system of nuclear energy, just as for fossil fuel plants, is the availability of the basic fuel source. In discussing uranium resources, ERDA utilizes some terms that require elucidation.

The first of these is "reserves," which "refers to the quantity of uranium in known deposits which it is calculated can be economically produced within the stated cost. The quantity, grade, and physical characteristics have been established with reasonable certainty by detailed sampling, usually by surface drilling initially, and later supplemented by underground drilling and sampling."[2]

There is then an overall category of "potential resources," which "are estimates of the quantity of uranium in addition to reserves that may exist in unexplored extensions of known deposits or in undiscovered deposits within or near known uranium areas. The estimates are based on extrapolations from explored to unexplored or incompletely studied areas applying favorability factors based on geologic evaluations."[3]

In another publication we find that "potential" is defined as follows:

---

*The concept of plutonium recycling entails extracting the 250 kilograms (kg.) of plutonium generated within the spent fuel rods of a typical 1,000-megawatt reactor and then converting this into new fuel rods that reduce the uranium demand. However, the extreme toxicity of plutonium, its 24,000-year radioactive half-life, and the fact that 11 kg. are sufficient to fashion a crude but effective atomic bomb have led the Nuclear Regulatory Commission (NRC) to impose a three-year moratorium on plutonium recycling. The NRC is studying to determine if effective safeguards for the handling and storage of plutonium can be developed at an economical cost. The White House Council on Environmental Quality has commented that the social, legal, and constitutional aspects that can affect the right of privacy and freedom of movement must also be examined. At this time, bills have already been introduced into Congress to prohibit plutonium recycling. For these reasons, it would appear that the plutonium question may be the clarion call of the environmentalists, and therefore plutonium recycle should not currently be counted upon in developing and utilizing forecasts.

TABLE 1.5

Uranium Requirements in Metric Tons of U_3O_8

	Low Growth No Pu Recycle		Moderate Low No Pu Recycle		Moderate High No Pu Recycle		High With Pu Recycle	
	Annual	Cumulative	Annual	Cumulative	Annual	Cumulative	Annual	Cumulative
1975	5,200	5,200	5,400	5,400	7,100	7,100	8,000	8,000
1976	9,400	14,600	9,900	15,300	10,200	17,300	12,100	20,100
1977	11,000	25,600	11,700	27,000	12,100	29,400	13,200	33,300
1978	12,600	38,200	13,700	40,700	15,500	44,900	18,200	51,500
1979	12,800	51,000	15,300	56,000	17,100	62,000	21,000	72,500
1980	15,600	66,600	17,700	73,700	19,900	81,900	22,800	95,300
1981	18,000	84,600	21,200	94,900	24,600	106,500	26,600	121,900
1982	21,900	106,500	25,200	120,100	29,900	136,400	33,000	154,900
1983	24,600	131,100	29,500	149,600	34,400	170,800	39,000	193,900
1984	29,000	160,100	32,800	182,400	38,700	209,500	44,400	238,300
1985	33,300	193,400	36,300	218,700	43,600	253,100	49,000	287,300
1986	36,600	230,000	44,000	262,700	47,000	300,100	53,200	340,500
1987	40,800	270,800	48,400	311,100	54,300	354,400	56,500	397,000
1988	44,600	315,400	52,700	363,800	61,100	415,500	63,200	460,200
1989	48,100	363,500	58,400	422,200	68,200	483,700	69,900	530,100
1990	52,100	415,600	63,200	485,200	74,500	558,200	80,600	610,700
1991	55,600	471,200	68,400	553,800	81,100	639,300	88,100	698,800
1992	59,700	530,900	73,600	627,400	87,900	727,200	96,400	795,200
1993	63,800	594,700	77,900	705,300	94,000	821,200	105,300	900,500
1994	66,300	661,000	82,300	787,600	101,100	922,300	114,700	1,015,200
1995	68,820	729,800	86,400	874,000	108,100	1,030,400	124,000	1,139,200
1996	71,600	801,400	90,700	964,700	114,900	1,145,300	135,200	1,274,400
1997	73,800	875,200	94,400	1,059,100	121,400	1,266,700	146,500	1,420,900
1998	75,500	950,700	97,600	1,156,700	127,100	1,393,800	156,400	1,577,300
1999	77,400	1,028,100	101,000	1,257,700	132,300	1,526,100	166,100	1,743,400
2000	78,600	1,106,700	103,200	1,360,900	134,100	1,660,200	173,700	1,917,100

Source: U.S. Energy Research and Development Administration, Office of Planning and Analysis, *Forecast of Nuclear Capacity, Separative Work, Uranium, and Related Quantities*, February 1975.

The potential uranium resources as estimated by the AEC are restricted to "conventional" uranium deposits and include uranium surmised to occur in (1) unexplored extensions of known deposits; (2) postulated deposits within known uranium areas; or (3) postulated deposits in other areas known to be geologically favorable for uranium.

These estimates of potential should not be interpreted as delimiting of the uranium resources; instead, they are appraisals of the amount of undiscovered uranium that is presumed to be present on the basis of rational geological evidence and/or exploration experience on hand at the time of estimation.[4]

ERDA has now subdivided the "potential" category into three classes: probable, possible, and speculative:

The "probable" potential resources are those estimated in known uranium districts, either in extensions of known deposits or in new deposits, within mineralization trends identified by exploration.

"Possible potential resources are in new deposits in formations or geologic settings known to be productive and which are located in productive geologic provinces.

The "speculative" potential resources are in postulated deposits, in formations, or geologic settings that have not been previously productive within a productive province or alternatively in new deposits postulated in geologic provinces that have not previously been productive.[5]

ERDA in various reports speaks about the reserves and potential available at various "cut-off" costs, usually at $8, $15, and $30 per pound. This is in fact a "marginal cost" figure, which means that if a mine is in operation already, the cut-off cost is equivalent to the cost of labor and operating expense of mining a given quality of ore. According to "Statistical Data of the Uranium Industry,"[6] this estimate does not take into account property acquisition costs, finding costs, and other past capital costs that are considered to be sunk costs. Neither are the cost of money, dollar reserves accumulated for replacing ore reserve, nor profit included.

In actual practice, these dollar amounts become meaningful only when they are used relative to one another as a method of evaluating the quality of different ore bodies. Therefore, a $15 cut-off reserve can be thought of as roughly twice as expensive to mine as an $8 reserve.

The total resources of uranium* up to the $30 cut-off figure at the various levels of reserves and potential are:[7]

---

*When uranium ore is mined, the uranium content is concentrated into a semi-refined uranium oxide U_3O_8 known as "yellowcake." Typically, over 500 tons of ore are necessary to produce one ton of yellowcake. The figures here and in Table 1.6 are for U_3O_8.

Known reserves	543,000 metric tons*
Plus potential: probable	1,034,000
Total	1,577,000
Plus potential: possible	1,215,000
Total	2,792,000
Plus potential: speculative	372,000
Total	3,164,000 metric tons†

COMPARISON OF URANIUM SUPPLY WITH DEMAND

There are two approaches for comparing uranium supply with demand:

1. Compare annual cumulative projected usage with supplies and then determine in what year supplies run out and all reactors would be without fuel.
2. Compute how many "life cycles" can be fueled with a given level uranium; that is, how many reactors could be supported for their lives of 30 years with the uranium available.

If the first approach is used and it is assumed that whatever reactors are built are fueled each year until the total supply of uranium is exhausted, a

---

*Patterson's figures were in short tons. The conversion formula for short tons to metric tons is:

Multiply by (2,000/2,204) or 0.90744101.

For metric tons to short tons,

Multiply by (2,204/2,000) or 1.102.

†"Uranium Reserves, Resources, and Production" citing R. D. Nininger (ERDA), Conference on Energy Development, Atomic Industrial Forum, February 1976 gives uranium reserves and resources as:

	MT U_3O_8
Known Reserves	581,000
Plus Potential: Probable	962,000
Total	1,543,000
Plus Potential: Possible & Speculative	1,689,000
Total	3,232,000

Plus another 127,000 MT may be recovered as a by-product of phosphate and copper production through the year 2000.

As can be seen, the overall rise in one year in all categories (including by-product production) is 3.5 percent. There has been a slight shift from "probable" to "known" but the two categories together have actually declined. A fair assessment would be that in one year the increase in uranium resources has not been significant.

TABLE 1.6

Metric Tons U3O8

	$8	$2 Increment	$10	$5 Increment	$15	$15 Increment	$30
Reserves	181,000	104,000	285,000*	95,000	380,000	163,000	543,000
Potential							
Probable	272,000	145,000	417,000	200,000	617,000	417,000	1,034,000
Possible	181,000	172,000	353,000	227,000	580,000	635,000	1,215,000
Speculative	27,000	73,000	100,000	91,000	191,000	181,000	372,000
Total	661,000	494,000	1,155,000	613,000	1,768,000	1,396,000	3,164,000

*An additional 82,000 tons U3O8 may be recovered as a by-product of phosphate and copper production through the year 2000.

Source: John A. Patterson, Chief, Supply Evaluation Branch, Division of Production and Materials Management, ERDA, *U.S. Uranium Situation*. Speech given at the Atomic Industrial Forum Fuel Cycle Conference 75, on March 20, 1975, data as of January 1, 1975. (Patterson's numbers were in short tons.)

comparison of the resources given in Table 1.6 with demand in Table 1.5 will show that known reserves are depleted in the following years:*

High case†	1990
Moderate high	1990
Moderate low	1991
Low case	1993

Since ERDA's projections do not extend beyond the year 2000, the years in which various levels of "potential" supplies would be exhausted cannot be determined. The early depletion of known uranium reserves is analogous to the oil situation where known domestic reserves are projected to be depleted around the turn of the century. As with uranium resources, additional potential oil may be discovered. However, due to the limits of current knowledge of all potential resources, oil-fired plants are no longer being built and existing ones are being converted to coal.

---

*Assumes spent fuel reprocessing capability reduces need for uranium.
†With Pu recycle.

With limited resources, a more cogent method of comparing supply and demand might be to calculate the number of reactors that can be fueled for their entire life cycles. For a 1,000 MWe reactor with a 30-year life, 5,069 metric tons (MT) are needed. This estimate is based on an initial charge of 545 MT and an annual reload of 156 MT (at a 65 percent capacity factor).[8] Dividing the tons of uranium available by 5,069 will yield the number of 1,000 MWe reactors that can be fueled for their lifetimes.*

It will be noted from Table 1.7 that, at most, 117 plants of 1,000-MWe capacity can be supported for their lifetimes by known domestic reserves. However, as of September 1975, the Nuclear Regulatory Commission had already

---

*At an efficiency rate for converting heat to electricity of 32.6 percent (known as 32.6 MWD_{th}/kg U), 1 kilogram (kg) of enriched uranium produces 258,200 kilowatt hours (kwh) or 0.2582×10^6 kwh. (MWD_{th} = thermal megawatt days.) Thus a reactor operating at 65 percent capacity would require 22,052 kg of enriched uranium per year.

(8760 hours/yr.) (1,000 MWe) (65% Cap. Fac.) =

$(8.76 \times 10^3) (10^6 \text{ KW}) (.65) = 5.694 \times 10^9$ KWH/yr.

$(5.694 \times 10^9)/(.2582 \times 10^6) = 22,052$ kg.

At 0.30 tails assay 7.08 kg. of natural uranium (U_3O_8) are required to produce 1 kg. enriched uranium.

1,000 kg = 1 MT

$(22,052 \times 7.08)/(1,000) = 156$ MT/ur. of natural uranium

Assuming a 30-year life for a plant, we have the following equation:

(initial load) + (years $\times$ reloads) = life cycle requirements for one plant.

545 MT + (29 $\times$ 156) = 5,069 MT

It should be noted that an efficiency rate of 32.6 percent is an assumed conversion rate utilized by ERDA in its computations, and our methodology has paralleled theirs. Whether or not this high an efficiency rate is justifiable is open to question. In the Bulletin of the Atomic Scientists of December 1975, M. C. Day ("Nuclear Energy: A Second Round of Questions") cites testimony of F. B. Baranowski of ERDA as testifying to Sen. Frank E. Moss on July 1, 1974, that the conversion ratio was 14 million kwh per short ton of uranium oxide, compared with the more commonly used value of 32 million kwh per short ton. Day does not draw any definitive conclusions, but if the efficiency rate is indeed less than 32.6 percent, the ability of uranium resources to sustain nuclear plants will be considerably less than indicated in this chapter which is based on ERDA assumptions.

TABLE 1.7

Reactors Supportable by Uranium Resources
65 Percent Capacity with no Plutonium Recycle

Resource Base	MT U$_3$O$_8$	Number of 1,000 MWe Plants Supportable	
		Without Reprocessing	With Reprocessing*
Known reserves	543,000	107	117
Plus potential:probable	1,034,000		
Total	1,577,000	311	341
Plus potential: possible	1,215,000		
Total	2,792,000	550	604
Plus potential: specu- lative	372,000		
Total	3,164,000	624	684

*If the reprocessing capability projected by ERDA does, in fact, turn out to provide 10 percent of the uranium needed, the life cycle demand is reduced to 4,620 MT. According to D. E. Saire (CONF-750209, pp. 67-72), reprocessing is expected to provide from 1.4 to 23 percent of total feed requirements between 1976 and 2000. In reality, reprocessing capability is far short of any projected targets, but an average of 10 percent is being utilized for discussion purposes.

Source: Compiled by the authors.

granted operating licenses to 37,000 MWe of nuclear capacity, construction permits for 65,000 MWe, and limited work authorization (ground breaking permits) for 18,000 MWe—a total of 120 GWe of committed nuclear capacity. Unless more "known reserves" are found, some of these reactors may have to shut down before their economic lives are completed. Prudence would dictate that no more new reactors be started until "potential resources" become additional "known reserves." However, ERDA projections show that by 1985, between 43 and 128 plants will have been built, which, based on known reserves, will not be able to function for their full economic lives. Table 1.8 illustrates this.

In fact, all projections forecast plants coming on stream *after* the known reserves would be fully committed (were producers to sign 30-year contracts). Under the moderate/high case (considered the "reference" case by ERDA) demand created by plants begun through 1990 would exceed all resources by 32 percent. It should be understood that the "potential resources" at this time—"probable," "possible," and "speculative"—are only "maybe's."

TABLE 1.8

Number of 1,000-Megawatt Plants Beyond Carrying Capacity of Fuel Supply Projected to Begin Construction*

	Low	Moderate/Low	Moderate/High	High
	Utilizing Known Reserves			
1975	43	68	88	128
1980	168	223	268	353
1985	328	428	523	673
1990	508	683	883	1,133
	Utilizing Known Reserves and Probable Potential Resources			
1980	—	—	44	129
1985	104	204	299	449
1990	284	459	659	909
	Utilizing Known Reserves, Probable and Possible Potential Resources			
1985	—	—	36	186
1990	21	196	396	646
	Utilizing Known Reserves and Probable, Possible, and Speculative Potential Resources			
1985	—	—	—	106
1990	—	116	316	566

*This table assumes 10 percent substitutability of reprocessed fuel for natural uranium. Dates shown are beginning project dates assuming a 10-year lead time to the start of commercial operation; i.e., Table 1.4 shows projected capacity in operation. This table relates to the *decision* time frame that occurs 10 years earlier.

Source: Compiled by the authors.

HOW PROBABLE IS SIGNIFICANT URANIUM EXPANSION?

If we judge the probabilities of significantly expanding uranium production from the reports of ERDA, we should not be too optimistic. First of all, the quantity of resources in the $15 to $30 cut-off range is a large question. We find that "a large portion is in the $15 to $30 cost increment for which data are not as well developed and mining experience is lacking."[9]

ERDA is expanding its search program significantly using aerial, surface, subsurface, and remote sensing techniques. By 1980, and Aerial Radiometric Reconnaissance Program should be completed. This will entail surveying the entire coterminous United States plus Alaska, flying over 700,000 miles at distances 200 to 800 feet above ground level.

Hope for significant finds should be tempered, however. Because of its radiological activity, uranium is considerably easier to detect than other minerals and most large deposits are likely to have been found. WASH 1242 states:

> Uranium deposits are relatively small compared to those of many
> other metals and the fossil fuels, and most near-surface deposits pre-
> sumably have been found. Thus, discovery of new deposits will
> become increasingly difficult and expensive. The resources that can
> be produced at a given cost are limited and cannot be expanded
> indefinitely.[10]

> Most uranium deposits are small, containing less than 100 tons
> of U_3O_8. The relatively few large deposits have the bulk of the
> reserves; 10 percent, or about 70 deposits, contain 85 percent of the
> $8 reserves.[11]

> The magnitude of the future exploration and production
> effort may be judged in terms of the size of the largest known
> uranium district in the United States, and one of the largest in the
> world, the Ambrosia Lake district in New Mexico. Past production
> and estimated remaining $8 resources in this district total about
> 150,000 tons of U_3O_8. The 1990 demand is equivalent to producing
> in one year 80 percent of the entire resources of Ambrosia Lake.
> The cumulative demand through 1990 (18 years) will require mining
> the equivalent of 6.5 Ambrosia Lake districts or four times the total
> United States production of 244,000 tons during the past 23
> years.[12]

According to WASH 1243, "The magnitude of the exploration and production effort ahead can be more clearly visualized by recognizing that the
demand through 2000 will require finding and mining out the equivalent of 16
Ambrosia Lake districts (New Mexico), the largest United States uranium district, or 10 times the total past production in this country over an equivalent
time span of about 25 years."[13]

Neither is drilling experience encouraging. "In the 1950s discovery rates
of more than 10 pounds of U_3O_8 per foot of drilling were achieved."[14] As
shown in Figure 1.2, discovery rates declined to five pounds per foot, and by
1972, had dropped even further to 1.8 pounds per foot. This means that over
five-and-one-half times as much drilling must be performed as was done in the
1950s for each increment of uranium found.

The magnitude of the required growth in uranium production can be
visualized by comparing projected U_3O_8 growth with the growth curves of other
commodities as shown in Figure 1.3. Another consideration is that as higher cost
uranium is found, the quality drops measurably. Eight dollar reserves contain ore
that is 0.20 to 0.22 percent uranium while $15 resources contain only 0.10 to
0.12 percent uranium, and $30 resources have only 0.07 percent uranium. Thus,
at $15, twice as much ore must be mined and processed to produce an equiva-
lent amount of uranium, and at $30, three times as much.

FIGURE 1.2

Exploration Activity—Reserves Additions

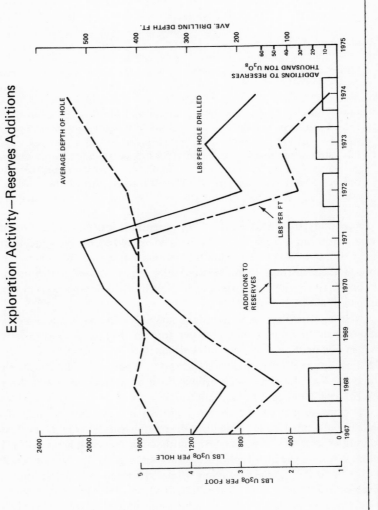

Source: Patterson, Figure 6. (See note 5.)

14

FIGURE 1.3

U.S. Annual Mineral Production
Growth Comparisons

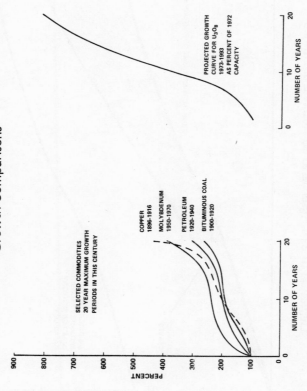

Source: WASH 1242 (May 1973), p. 4.

15

FIGURE 1.4

Depletion of $15 Reserves as $8 Reserves are Mined

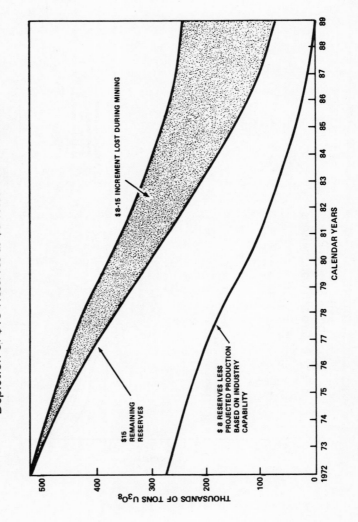

Source: WASH 1242, p. 8.

The problem of having adequate resources is further compounded by the fact that a major proportion of higher cost resources are in the same deposits as the lower cost ore. As long as $8 ore is mined, higher cost material is bypassed and much of it will become lost entirely or become even more expensive for later recovery. This can be seen in Figure 1.4.

Occasionally, the potential for mining various types of shale is cited as a source of uranium. Considering that the uranium content of the highest quality shale is 60 to 80 ppm (parts per million) or 0.006 to 0.008 percent compared with 0.20 to 0.22 percent for ore classified as an $8 resource, it can be seen that 30 times greater quantities of ore would have to be mined than currently.[15]

FOREIGN RESOURCES

Based on data available, looking to foreign sources would be hazardous, and would endanger the economic security of the United States more than does importing foreign oil. WASH 1242 notes, "The foreign supply-demand situation is expected to be much like that projected for the United States with no excess of supply unless major new discoveries are made."[16] (See Figure 1.5.)

Regarding foreign uranium, Patterson notes, "Foreign resources are somewhat larger than U.S. resources. Production capability, however, is similar to U.S. capability, and there are limitations on the degree to which foreign capability can be expanded. South African uranium production is limited by rates of production of gold ores. Canadian resources are contained in just a few mines. Furthermore, some foreign government policies restrict development, exploitation, and export of their resources."[17]

Furthermore, foreign resources are constricted by political decisions of other governments. Forbes, the financial biweekly, stated, "The Australian government put a halt to all uranium exports in 1971, and doesn't seem to be in much of a hurry to change that policy."[18] Friends of the Earth confirmed this by reporting that the Australian labor union that mines the ore is fundamentally opposed to mining any further ore, and the Australian Atomic Energy Commission conceivably will not permit foreign exploration after 1977.

Regarding other foreign free-world sources of uranium, Forbes headlined its article with an ominous prediction: "In 1977, U.S. utilities will start importing uranium to fuel nuclear power stations. When they do, they may be at the mercy of a uranium OPEC." Forbes continued, "Control of the uranium cartel may be far more concentrated than would appear at first glance."[19] Except for Australian ore, virtually all foreign free-world mines are controlled by one family, the Rothschilds of France and England. The price control that they exert is evidenced by the fact that foreign producers would not even accept bids in January 1975 for future delivery at prices of $27 a pound, which was three to four times the price then being paid domestically under contract.

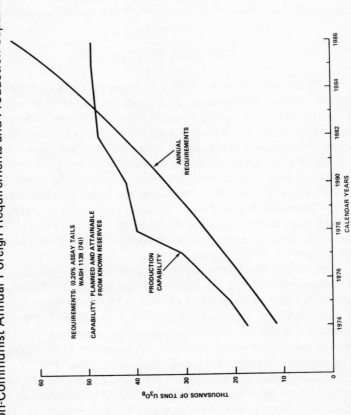

FIGURE 1.5

Non-Communist Annual Foreign Requirements and Production Capability

REQUIREMENTS: 0.20% ASSAY TAILS
WASH 1139 (74))

CAPABILITY: PLANNED AND ATTAINABLE
FROM KNOWN RESERVES

ANNUAL
REQUIREMENTS

PRODUCTION
CAPABILITY

THOUSANDS OF TONS U3O8

CALENDAR YEARS

1974   1976   1978   1980   1982   1984   1986

Source: John A. Patterson, speech given to the 17th Minerals Symposium of the American Institute of Mining, Metallurgical, and Petroleum Engineers at Casper, Wyoming, May 11, 1974, figure 12.

18

TABLE 1.9

Indicators of Trend of Future U.S. Uranium Resource Development

Positive Substantially Larger Resources	*Negative* Minor Additional or Inadequate Resources
Experience is that mineral resources ultimately prove larger than current estimates.	Exploration and appraisal effort more intensive than for other minerals; future growth may have been anticipated by present potential estimates.
Substantial portion of United States incompletely explored.	Some exploration has been done over most of United States.
Exploration continues to add to resources.	Exploration success per unit of effort has been decreasing. No new major districts found in 17 years.
Radioactivity simplifies exploration.	Increases likelihood that best resources are already found.
Known reserves are in a few small areas.	Resources may be restricted to known areas for geologic reasons. United States deficient in several minerals.
Large low grade but economic deposits exist elsewhere in the world.	Similar geologic environments in United States are limited.
Nuclear power costs insensitive to higher uranium prices. Price increases will stimulate exploration.	Time frame short–30-50 years. Long lead times, competition for capital, market uncertainties, and inflation delay action.
Large areas of the world have not been explored.	Political considerations limit foreign exploration and restrict access to foreign supplies.

Source: Patterson, Fig. 10 (see note 5).

On July 7, 1976, the *Wall Street Journal* reported that the Justice Department was investigating the uranium industry in a wide-ranging hunt for a price-fixing conspiracy. The article also highlighted shortages already appearing with the report that: "Late last year the South Texas Project, which includes Houston Lighting & Power Co. and municipal utilities for San Antonio and Austin, sought bids for uranium to fuel a big nuclear plant. They didn't receive any bids that met specifications."[20] The positive and negative indications regarding the prospects for uranium resource development are summarized in Table 1.9.

SUMMARY

Based upon forecasts of projected energy and nuclear plant growth by ERDA, the uranium supply situation from 1985 to the end of the century will become at least as critical as the availability of domestic oil. Based on ERDA projections, all known domestic·uranium reserves will be exhausted between 1990 and 1993. Taking another viewpoint, domestic reserves can support 107,000 megawatts (MWe) of nuclear plants for their lifetime of 30 years; however, as of September 1975, the Nuclear Regulatory Commission (NRC) reported that 120,000 MWe had already been granted operating licenses, construction permits, or limited work authorizations. Unless more uranium is found, some of these plants may have to shut down before their economic lives are completed.*

CONCLUSION

The actual sources of uranium may be greater than those indicated. On the other hand, a high percentage of the "maybe's" represented by the "probable," "possible," and "speculative" may turn out to be considerably smaller than anticipated. The forecasts may actually be conservative in even more respects because the requirements for uranium are based on the assumption that by 1983, approximately 9 percent of the fuel needs will be met through reprocessing; at this time, no commercial reprocessing plant is operating, and all prior attempts have resulted in failures.

---

*On March 19, 1976, the *Wall Street Journal* reported: "Westinghouse noted [in its annual report] that it is defending 17 lawsuits filed by 27 public utility customers alleging breach of the uranium supply contracts after the company's assertion in September that it was excused from fulfilling the pacts because it wouldn't be commercially practical. Westinghouse repeated statements that it is about 65 million pounds short of uranium to fulfill the disputed contracts, which call for deliveries over the next 20 years; the average price under the pacts is $9.50 a pound, plus certain escalation, compared with recent market quotations of about $40 a pound."

To rely upon foreign sources for fueling an industry that will have hundreds of billions of dollars tied up in capital-intensive nuclear plants would leave the United States open to exorbitant blackmail and would pose an unacceptable level of danger to the nation's security. The experiences with the oil embargo of 1973, the gradual phasing out of Canadian oil, and the continuing revolutions within the underdeveloped countries (which are often suppliers of vital raw materials) would seem to indicate that reliance upon foreign nations for vital substances is an unwise policy. A prudent course of action would be to build only that number of nuclear reactors for which fuel supplies can be contracted with reasonable certainty over the economic lives of the plants.

NOTES

1. U.S. Energy Research and Development Administration (ERDA), Office of the Assistant Administrator for Planning and Analysis, *Total Energy, Electric Energy, and Nuclear Power Projections*, February 1975.

2. U.S. Atomic Energy Commission (AEC), *Nuclear Fuel Resources and Requirements*, WASH 1243 (April 1973): 12.

3. ERDA, Office of Planning and Analysis, *Forecast of Nuclear Capacity, Separative Work, Uranium, and Related Quantities*, February 1975.

4. AEC, *Statistical Data of the Uranium Industry*, GJO 650-100 (74), (January 1, 1974): 25.

5. John A. Patterson, Chief, Supply Evaluation Branch, Division of Production and Materials Management, ERDA, *U.S. Uranium Situation*. Speech given at the Atomic Industrial Forum Fuel Cycle Conference 75, March 20, 1975, p. 6.

6. GJO 100 (74), op. cit., p. 13.

7. Patterson, op. cit.

8. AEC, *Nuclear Fuel Supply*, WASH 1242, May 1973, p. 2.

9. Patterson, op. cit., p. 7.

10. WASH 1242, op. cit., p. 8.

11. Ibid., p. 5.

12. Ibid., p. 7.

13. WASH 1243, op. cit., p. 21.

14. Ibid., p. 23.

15. WASH 1242, op. cit., p. 16.

16. Ibid., p. 11.

17. John A. Patterson, speech given to the 17th Minerals Symposium of the American Institute of Mining, Metallurgical, and Petroleum Engineers at Casper, Wyoming, May 11, 1974, p. 8.

18. *Forbes*, "It Worked for the Arabs . . . ," January 15, 1975, p. 19.

19. Ibid.

20. The *Wall Street Journal*, "Justice Unit's Study of Uranium Industry Turns into Hunt for Price-Fixing Scheme," July 7, 1976.

2

The first step in the nuclear fuel cycle, the availability of uranium ore, has already been discussed. But as with any business analysis, the total system must be fully evaluated before any definitive conclusion can be reached.

The nuclear fuel cycle consists of a sequence of eight steps, each of which is integral and necessary to the proper and economic functioning of the entire system.* These steps are:

1. mining and milling of uranium
2. conversion of ore concentrate to uranium hexafluoride (UF_6)
3. enrichment in the isotope uranium-235 (U^{235})
4. conversion of enriched uranium to fuel material, UO_2
5. fabrication of fuel elements
6. use of fuel elements in nuclear power plants
7. reprocessing of spent fuel
8. disposal of radioactive waste

Using these eight steps as the framework within which to discuss the economics of nuclear fuel, we should seek to answer these questions:

---

*This is the cycle for light water reactors, which comprise the vast majority of U.S. nuclear power generators.

A. Is the *entire* system comprising the nuclear fuel cycle reliable? To what degree?
B. What should our confidence level be in the system's operational capability?
C. Is nuclear power a sound (or attractive) commercial investment at each stage?
D. Will the growth of nuclear power contribute to the economic and physical security and stability of the United States?

MINING AND MILLING

The supply and demand aspects of uranium mining have been discussed so far as availability of resources, but the availability of capital equipment and the capability of mining the ore are also necessary to insure a sufficient continuous flow of fuel, just as oil rigs and pipelines are necessary to provide a flow of petroleum.

Table 2.1 shows a daily processing capability within the United States of 28,550 tons of ore. This equates to 57.1 tons of "yellowcake" or U_3O_8 per day, or 18,912 MT per year. Based on the low growth projection of uranium needs, this capability will be adequate through 1981; for the moderate low growth case, through 1980; and for the moderate high growth case, through 1979.

It would appear that five to eight years is ample time for developing new mines, but this is really only just enough. At this time, there appear to be no announced plans for any significant additional mine openings. If ore production capability is to maintain pace with uranium demand, much larger capital investments than now programmed will have to be allocated to uranium mining. Frank P. Baranowski of ERDA stated in November 1974:

> The U.S. will need to build roughly 12-14 additional mine and mill complexes over the next five years or so, assuming no imports. If the maximum allowable imports are assumed, the number of additional mine and mill complexes will still be about six to seven over the same period. These mills are each equivalent to 1,000 tons U_3O_8 per year capacity, which is the average for the currently installed U.S. production capability. . . .
>
> Further, any delay in time of installation of reprocessing plants or enrichment plants or reduction in plant performance would require additional feed and therefore more mills to keep the enriched uranium in the fuel cycle in balance. . . .
>
> It is clear that a major expansion of effort and capital investment and, I might add, luck are needed to bring in more uranium on a timely basis. AEC stockpiles and the capability built for U.S. defense needs will soon be exhausted and industry must expand its capability soon to meet the demand.[1]

TABLE 2.1

Ore Processing Capacity of Uranium Milling Companies and Plants

Company	Plant location	Nominal capacity tons ore per day
The Anaconda Company	Grants, New Mexico	3,000
Atlas Corporation	Moab, Utah	1,500
Continental Oil Company	Falls City, Texas	1,750
Cotter Corporation	Canon City, Colorado	450
Dawn Mining Company	Ford, Washington	500
Federal-American Partners	Gas Hills, Wyoming	950
Exxon Nuclear Company, Inc.	Powder River Basin, Wyoming	2,000
Kerr-McGee Nuclear Corporation	Grants, New Mexico	7,000
Petrotomics Company	Shirley Basin, Wyoming	1,500[a]
Rio Algom Corporation	La Sal, Utah	500
Union Carbide Corporation	Uravan, Colorado	1,300
Union Carbide Corporation	Natrona County, Wyoming	1,000
United Nuclear—Homestake Partners	Grants, New Mexico	3,500
Utah International, Inc.	Gas Hills, Wyoming	1,200
Utah International, Inc.	Shirley Basin, Wyoming	1,200
Western Nuclear, Inc.	Jeffrey City, Wyoming	1,200[b]
Total		28,550

[a]Currently closed.
[b]Currently closed for modifications.
Source: U.S. Atomic Energy Commission, *The Nuclear Industry*, WASH 1174-74, (1974): 45.

Since recent data does not indicate that new mines are being brought on stream to fill the mining shortfall that could develop between 1978 and 1981, there could be a serious bottleneck within only a few years. Even if new plants were announced, shortages of U_3O_8 are highly possible as early as 1978.

As can be seen from Figure 2.1, by 1980, enrichment contracts with ERDA call for more uranium than is projected to be available from mining production. If production capability were expanded over the next five years, by 1981 lower grade $15 resources could be mined, thus increasing output as shown in Figure 2.2. However, this would necessitate increasing the current maximum possible throughput of current mills from about 28,000 tons of ore per day to

FIGURE 2.1

Estimated Schedule of Maximum Reasonable Production Capability from $8 Resources Compared with U.S. Separative Work Commitments

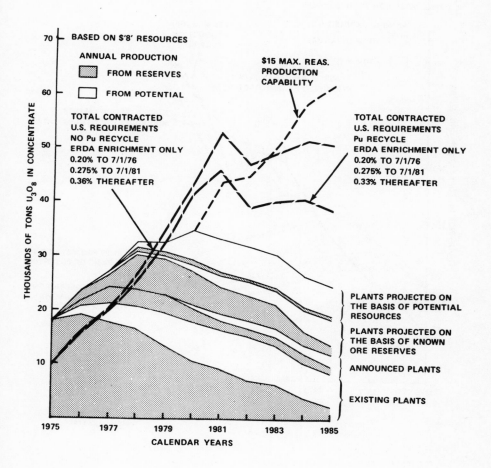

Source: CONF 750209, Figure 16, p. 100.

FIGURE 2.2

Projected Maximum Reasonable ("Could Do") U3O8 Production Capability of the Uranium Mining Milling Industry Through 1985

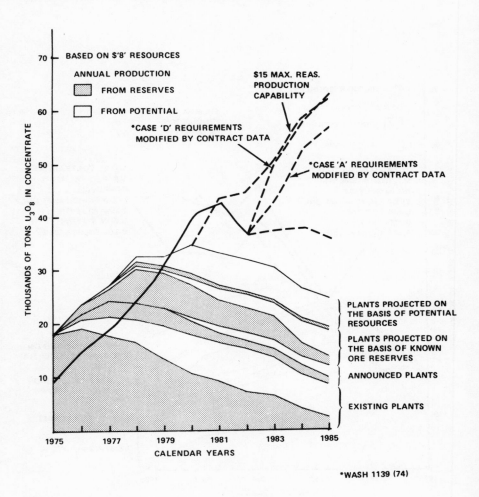

*WASH 1139 (74)

*Case "A" and "D" requirements lines are shown in these charts. In WASH 1139 (74), Case "A" was the lowest forecast of nuclear capacity and case "D" assumed a general reduction in the growth of electricity use. These forecasts were subsequently updated to the ERDA forecasts used throughout this report.

Source: CONF 750209, Figure 6, p. 89.

FIGURE 2.3

$8 Production Capability Scheduled to Meet Domestic U3O8 Requirements

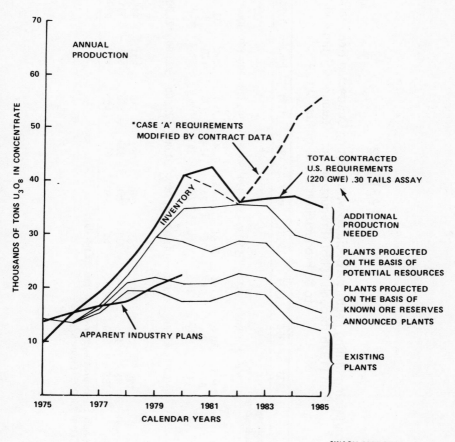

*WASH 1139 (74)

Source: CONF 750209, Figure 7, p. 90.

27

TABLE 2.2

Exploration/Mining/Milling Factors Associated with Meeting Contracted Requirements for the Operating Plan (1975 through 1985) (Production from Resources Estimated 1/1/75 to be Producible at $8/lb. or Less U3O8)

	1975	1976	1977	1978	1979	1980	1981	1982	1983	1984	1985	Total
Exploration (Including 8-Year Forward Reserves)												
Number rigs	170-250	250-330	300-370	250-330	300-370	300-370	300-370	300-370	300-370	250-330	250-330	
Feet of drilling (millions)	20-25	30-40	35-45	30-40	35-45	35-45	35-45	35-45	35-45	30-40	30-40	350-455
*Mines**	37	43	54	73	96	114	125	120	120	120	120	
Mills	14	14	14	16	20	26	30	30	30	30	30	
Manpower (manshifts)	6,000	8,000	10,000	14,000	18,000	21,000	22,000	22,000	22,000	22,000	22,000	
Costs (millions of dollars)												
Capital costs	170	225	265	265	290	295	310	325	335	315	315	3,110
Operating costs	85	115	140	200	250	320	350	335	335	335	335	2,800
Total costs	225	340	405	465	540	615	660	660	670	650	650	5,910

*Includes only mines producing over 100 tons ore per day.
Source: CONF 750209, Figure 8, p. 92.

28

TABLE 2.3

Exploration/Mining/Milling Factors Associated with Production to Fill Contracted Requirements, Operating Plan Through 1982, and Case A Growth Requirements (0.30 Percent Tails Assay) for 1983 Through 1985 (Production from Resource Estimates 1/1/75 to be Producible at $8/lb. or Less U3O8)

	1975	1976	1977	1978	1979	1980	1981	1982	1983	1984	1985	Total
Exploration (Including 8-year forward reserves)												
Number rigs	250-330	330-415	460-580	500-630	670-830	710-880	750-920	830-1,010	880-1,080	920-1,170	960-1,210	
Feet of drilling (millions)	30-40	40-50	55-70	60-75	80-100	85-105	90-110	100-125	105-130	110-140	115-145	870-1,080
Mines *	37	43	54	73	96	114	125	125	137	167	186	
Mills	14	14	15	16	20	26	30	30	34	42	46	
Manpower (manshifts)	6,000	8,000	10,000	14,000	18,000	21,000	22,000	24,000	25,000	31,000	34,000	
Costs (millions of dollars)												
Capital costs	255	350	415	425	435	455	485	580	605	800	835	5,640
Operating costs	85	115	140	200	250	320	350	360	385	445	490	3,140
Total costs	340	465	555	625	685	775	835	940	990	1,245	1,325	8,780

*Includes only mines producing over 100 tons ore per day.
Source: CONF 750209, Figure 9, p. 93.

29

55,000 tons per day. For the period up to 1982, uranium feed deficits can be made up from government stockpiles as shown in Figure 2.3. The dark line labeled "Apparent Industry Plans" reflects information available to ERDA in late 1974.

Paul C. de Vergie of ERDA stated, "An estimate of the factors associated with annual production to fill the contracted requirements through 1985 (Table 2.2) indicates that a minimum level of about 35 to 45 million feet of drilling would have to be maintained, up to 30 mills and 120 large mines would be needed, and capital costs of $300 million to $325 million per year would be involved in the years after 1980."[2] This compares with 14 mills and 37 mines (producing over 100 tons of ore per day) for 1975. Table 2.3 shows that the drilling rate will have to triple by 1980, and quadruple by 1984. To say the least, the projected uranium requirements are very demanding.

CONVERSION TO UF_6

The next stage in the fuel cycle is the conversion of ore concentrates into UF_6 gas (uranium hexafluoride). As of 1974, private industry had an annual 17,000 metric ton capacity. Currently, Kerr-McGee Corporation plans to begin the doubling of its facilities from 4,500 MT to 9,000 MT, thus providing a capacity of 21,500 MT by 1977 or 1978. The projected cost of the plant addition in 1974 was $7 million. Even if the cost were double, the small size of the expenditure would appear to indicate that this is a technologically simple process. Thus, this stage in the fuel cycle would appear to present no future problems.

ENRICHMENT

By contrast, the enrichment phase has been one of the costliest and most difficult to put on a commercial basis. In this step, the UF_6, which contains uranium with a U^{235} content of 0.71 percent, is raised to a 3.21 percent content level. In the United States, enrichment is accomplished by gaseous diffusion plants in which the gaseous UF_6 is pumped through a series of porous barriers. Each barrier discriminates against the heavier U^{238} isotope by a theoretical maximum factor of 1.0043. The UF_6 must pass through 1,200 to 1,700 barriers before the uranium is 4 percent U^{235}. The first diffusion plant was constructed at Oak Ridge, Tennessee in 1943. The other two existing plants at Paducah, Kentucky and Portsmouth, Ohio, were built by 1955. All three plants are government-owned but are operated by private corporations.

In order to follow a discussion of the enrichment process, an understanding of the four key elements in the process is helpful. They are:

1. *Normal feed.* This is natural uranium with a U^{235} content of 0.711 percent.
2. *Enriched product.* This is uranium with a higher content of U^{235}, normally 3.21 percent. In technical parlance, this is referred to as having a "product assay of 3.21 percent."
3. *Tails.* This is the by-product or waste product of the process, which has a "tails assay" normally between 0.20 percent and 0.30 percent.
4. *Separative work unit* (SWU). This is merely a measure of work expressed in kilograms, just as a foot is a measure of length or a pound is a measure of weight.

In reading ERDA forecasts, it will be observed that they are usually based upon a "tails assay"—sometimes 0.20 percent, other times 0.30 percent, and occasionally values in between or even higher. Assuming that the "feed" is uniformly 0.711 percent U^{235}, the tails assay chosen will be a determinant of: 1) the amount of feed required, and 2) the amount of separative work done. There is a trade-off between work performed and feed required if the end product is to be the same. If less work is performed, more feed is required. If more work is performed, less feed is required.

Let us draw an analogy. You're making a glass of fresh orange juice. The harder you squeeze each orange, the more work you do, but fewer oranges are required than if you don't squeeze out every last drop. Thus, the difference between a tails assay of 0.20 percent and 0.30 percent is that at the lower level, more work has been done to extract the valuable U^{235} from the natural uranium. Therefore, at 0.20 tails assay, only 5.9 kg. of natural uranium feed is required for each kilogram of enriched uranium, whereas 7.08 kg. of feed is necessary at the 0.30 level.

A logical question that may arise at this point would be, "Then why not produce only at the 0.20 tails assay level? " The reason is quite simple. At the 0.30 tails level, the same facility can produce more enriched uranium in a year than at the lower level, and every kilogram that can be produced will be needed.*

The ERDA enrichment complex is scheduled to have a 27.6 million SWU capacity when all plant improvements are completed in 1979. Based on ERDA commitments as shown in Table 2.4 and Figure 2.5, there is now an inventory cushion of enriched uranium, and it should grow over the next few years. However, this forecast is contingent upon a number of events occurring on schedule.

---

*Fortunately, however, these tails are not lost. They are stored as solid UF_6 in 14-ton capacity cylinders at the plants for future use; when uranium stocks begin to be depleted, the waste can be reused and much of the remaining U^{235} extracted. (Of course, storing hundreds of thousands of tons in cylinders is not a minor or inexpensive undertaking.)

FIGURE 2.4

Separative Work Example

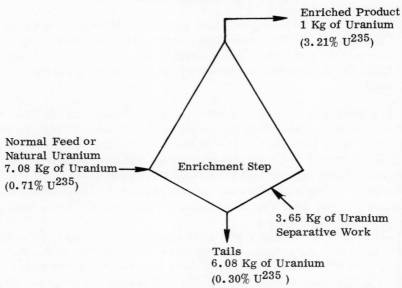

Enriched Product
1 Kg of Uranium
$(3.21\% \ U^{235})$

Normal Feed or
Natural Uranium
7.08 Kg of Uranium
$(0.71\% \ U^{235})$

Enrichment Step

3.65 Kg of Uranium
Separative Work

Tails
6.08 Kg of Uranium
$(0.30\% \ U^{235})$

Source: U.S. Atomic Energy Commission, *AEC Gaseous Diffusion Plant Operations*, ORO-684, Appendix 3, January 1972.

TABLE 2.4

Separative Work at 0.30 Tails (millions of SWU)

Fiscal Year	Production	Commitments	Net	Cumulative Net
Beginning Inventory				14.2
1975	12.4	7.4	5.0	19.2
1976	15.4	8.3	7.1	26.3
1977	17.5	12.5	5.0	31.3
1978	19.4	12.6	6.8	38.1
1979	21.3	17.4	3.9	42.0
1980	23.8	23.5	0.3	42.3
1981	25.3	28.2	(2.4)	39.9
1982	25.3	30.7	(5.4)	34.5
1983	25.3	26.0	(0.7)	33.8
1984	25.8	28.4	(2.6)	31.2
1985	27.6	29.9	(2.3)	28.9
1986	27.7	28.9	(1.2)	27.7

Source: U.S. ERDA, CONF-750209, derived from data presented on p. 53.

FIGURE 2.5

Annual Separative Work Production Capabilities versus Contracted Commitments

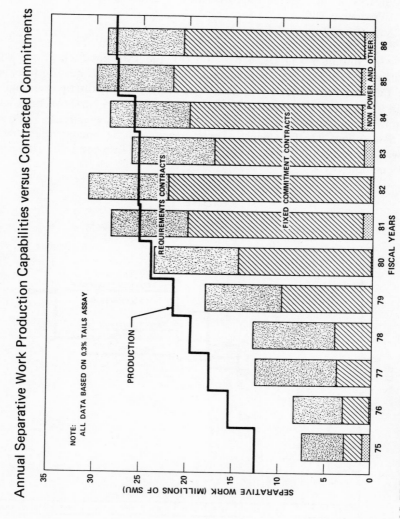

Source: U.S. ERDA, CONF-750209, p. 63.

33

Those conditions that could set this timetable back and affect the supplies of enriched fuel include:

1) potential power shortages from the Tennessee Valley Authority (TVA), the source of power for the enrichment plants
2) shortages of feed material
3) slippages in the Cascade Improvement Program (CIP) and Cascade Uprating Program (CUP)
4) "Acts of God"

In November and December of 1974, the TVA cut back power which resulted in 750,000 SWU of production being lost.[3] ERDA has contracts with the TVA for delivery of power. However, delivery of the total power commitment is not firm (100 percent available) until 1984. Figure 2.6 shows three levels of power, the power level currently being used for planning purposes (indicated by dotted line), the firm power commitment, and the amount of contracted power. The overall difference in production between firm and contracted power is 19 million SWUs.

FIGURE 2.6

3-Site Average Annual Power Levels

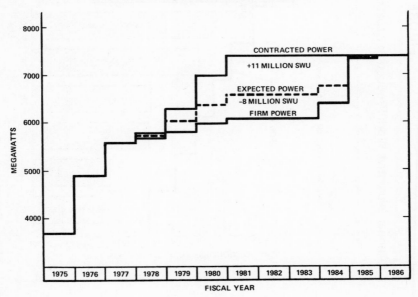

Source: U.S. ERDA, CONF-750209, p. 42.

Earlier in this book the potential shortage of U_3O_8 was presented. Prompt investments in new mines by corporations could conceivably pre-empt a shortage of uranium feed, at least in the near future. While new mines were not attractive at uranium prices of $8 per pound, capital investment in this sector should prove to be profitable at $24 or more. Nonetheless, a shortage of feed material is now anticipated.

CIP and CUP are described by ERDA as follows:

> The Cascade Improvement Program is a program of installing new and improved technology through equipment modifications such as rebuilt and improved compressors, piping, and control valves with better aerodynamic characteristics, new and improved barrier, etc. The Cascade Improvement Program is a capital program which will enlarge the capacity of the three facilities through equipment change-out and without significantly changing operating costs since no additional power is involved.
>
> The Cascade Uprating Program involves refurbishment and additional capacity of the electrical and utilities systems . . . and although it is a capital improvement program will require the purchase of additional power to take advantage of the improved facilities, and as a result, will affect operating costs.[4]
>
> While the CIP and CUP construction is proceeding essentially on schedule, there is always the possibility that these programs could slip and result in less than expected separative work production. After all, it is a billion dollar program with closely integrated scheduling and stretched out over many years. Perturbations in this program might come about by 1) mismatches of funding and construction, 2) delays in the procurement of critical items, and 3) in strikes at either our plants or the vendors' plants supplying critical components. For example, a one year slippage in these programs could result in as much as 10 million SWU less production. While we have great confidence in our development efforts, less separative work could result from lower than expected performance gains in the CIP and CUP. Not all of the technology has been proven yet but we have been attaining our projected levels of improved performance so far and have confidence that this will continue to be the case. We feel that any variance from expected performance gains will be relatively small.[5]

Regarding "Acts of God": in 1974, a switchyard at one of the generating plants supplying power to Portsmouth, Ohio was hit by a tornado. Coupled with TVA reductions, 5 million SWU was lost.[6] While the probability of "acts of God" occurring is low, they clearly can happen and would slow down production.

Because the construction of nuclear reactors is currently falling behind schedule even with minor slippages, ERDA may be able to fulfill its contracts.

However, ERDA is signing no more enrichment contracts and those utilities without a firm ERDA commitment may find themselves without fuel. Because 37 percent of the ERDA output is committed to foreign buyers, only 17.4 million SWU of the plant output is available to meet domestic needs. Yet, the requirements forecasts show that at a 0.30 percent tails assay, this capacity level will be reached by the following years:

Low Case	*Moderate/Low*	*Moderate/High*	*High*
1984	1984	1983	1982

Theoretically, an enrichment plant can be built within eight years, but no large commercial plant has ever been built from scratch, and to count on a new commercial venture meeting an initial timetable would appear to be naive. Even if a gaseous diffusion facility could be built in eight years, the power plants required to provide electricity to the enrichment plant require a 10-year lead time if they are nuclear.

FIGURE 2.7

Typical Lead Times in Enriched Uranium Supply

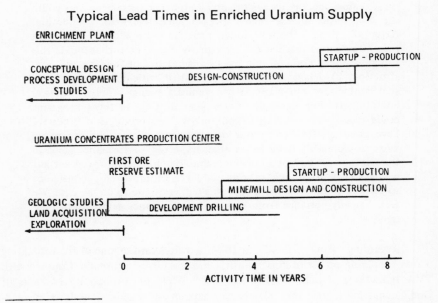

Source: U.S. AEC, *Nuclear Fuel Resources and Requirements*, WASH 1243, (April 1973): 47.

With an eight- to 10-year lead time required for an enrichment plant (see Figure 2.7) unless there is considerable slippage in reactor construction from the ERDA forecasts, there will be inadequate supplies of fuel for many of the plants being built. The government would then be caught in the dilemma of allowing domestic reactors to be closed down or of reneging on foreign contracts with a concomitant deterioration in foreign relations. WASH 1243 states, "In the future the utility industry should no longer expect a readily available supply of enriched uranium because of the long lead times involved in installing the necessary new facilities and because large expenditures must be supported by firm supply commitments to minimize the suppliers' dollar risks."[7]

The government no longer wants to build these plants, and to date, no private corporation has committed itself to the risk of a $3.27 billion investment (1975 dollars) estimated for a 9.0 million SWU plant.* Not only is this expenditure required for the enrichment facility, but a utility must commit to supply about 2,000 MWe of electricity to the enrichment plant, at a cost, in 1975 dollars, of $1.3 billion.

The problems of developing sufficient enrichment capacity go beyond that of merely coping with adquate lead times. The major problem is the difficulty of raising the amount of capital that would be required to build adequate enrichment facilities between now and the year 2000. To comprehend the magnitude of the problem, the SWU requirements together with the number of 9.0 million SWU plants needed and the capital costs are set forth in Table 2.5.

These capital requirements are a tall order. John La Fond of ERDA says, "Private financing of the magnitude necessary for a private enrichment plant is just not available in the classic commercial sense. There are just too many other things that are a better bet so far as risk is concerned."[8] This is evidenced by the extended hearings that Congress has been holding regarding the proposal that the government underwrite the proposed plant to be built by Uranium Enrichment Associates (UEA), a consortium of Bechtel Corp. and Goodyear Tire and Rubber Company. The reluctance of private industry to undertake the project without guarantees clearly indicates the magnitude of the risks involved.

NUCLEAR FUEL PROCESSING AND FABRICATION

The enriched fuel (UF_6) is converted to UO_2, which is then formed into pellets and sintered to achieve the desired density. The finished pellets are loaded into Zircaloy or stainless steel tubes, fitted with end caps, and welded. The completed fuel rods are assembled in fixed arrays to be handled as fuel elements. There are ten plants in the United States owned by corporations that

---

*As will be seen in Chapter 6, a cost of $7.23 billion in 1975 dollars will probably be more accurate.

TABLE 2.5

Enrichment Capacity and Capital Requirements at 0.30 Tails Assay (no Pu Recycle)

	Low Case	Moderate/ Low	Moderate/ High	High
Total U.S. capacity needs for year 2000 (millions of SWU)	48.8	63.7	82.6	106.2
Less ERDA capacity	(17.4)	(17.4)	(17.4)	(17.4)
Capacity to be added	31.4	46.3	65.2	88.8
No. of 9.0 SWU plants required	3.48	5.14	7.24	9.87
1975 dollars (in billions)*	15.7 to 29.7	23.1 to 43.8	32.6 to 61.8	44.4 to 84.2

*Includes costs of electric support power plants. This spread in costs based on a range of estimates of capital costs per SWU.

Source: Enrichment capacities are extracted from *Forecast of Nuclear Capacity, Separative Work, Uranium, and Related Quantities, United States, February 1975*, ERDA, Energy Systems Analysis Branch, Office of Planning and Analysis.

perform this activity. In 1974, a plant to perform fuel fabrication for 50 reactors of 1,000 MWe capacity was estimated to cost between $80 and $100 million.[9] This step in the nuclear fuel cycle appears to be relatively simple, and other than the increasing costs of labor and plant construction, it should pose no problems in the future.

REPROCESSING OF SPENT FUEL

Whereas fuel from fossil plants discharges ash with no residual fuel content, fuel discharged from nuclear reactors contains sizable quantities of "unburned" U^{235} and plutonium. Since the content of the spent fuel still has considerable value, the fuel should be reprocessed and usable uranium recovered, to be fed into the enrichment plant along with fresh feed. There are also plants to recover the plutonium that has been formed, but as discussed in Chapter 1, this process is now in limbo.

Although fuel reprocessing is considered to be an integral link in the nuclear fuel cycle, at the present time no commercial spent fuel processing facility is operating in the United States. The only one that has operated in the past was a 350 MT per year plant operated by Nuclear Fuel Services (NFS) in West Valley, New York, from 1966 to 1972. However, due to delivered dosages of radiation to workers that were 17 times those permitted by the Nuclear Regulatory Commission standards,[10] the plant is now closed and is undergoing modernization and expansion with a scheduled reopening in 1979, or later, at a capacity of 750 MT per year.

General Electric spent six years and $64 million utilizing a new untested technology on what was to be the Midwest Fuel Recovery Plant at Morris, Illinois. In 1968, GE broke ground and projected a completion date of 1970, a capacity rate of 300 MT per year, and a total cost of $36 million. By mid-1974, the plant was uncompleted and abandoned as a reprocessing facility. ". . . Plumbing problems had been expected, but the 'special design' worked out to solve these problems had only exacerbated them. Moreover, equipment which would be so radioactive under normal operating conditions that it could never again be touched by human hands turned out to be repairable only by human hands. The study [by GE] found that the Morris plant might be able to process 50 to 100 tons of fuel a year, but in the likely event of a major breakdown it could be closed for 'years'."[11]

Allied-General Nuclear Services (AGNS), a partnership between Allied Chemical and the General Atomic Company, is constructing a plant at Barnwell, South Carolina. This plant has a design capacity of 1,500 MT per year, and is scheduled to open in 1977. (If the usual past experience of nuclear facilities applies to Barnwell, the plant will open later than 1977 and will have a capacity less than 1,500 tons.) In mid-1974, when the plant was scheduled to open in 1976, the AEC wrote, "Many observes have expressed doubt that the AGNS and NFS plants will start up on, or even close to, schedule and that the plants will be able to meet and sustain their design capabilities. They point out that there have already been several slippages in projected start-up dates for both plants, and that reprocessing experience suggests that achieving current projects will be extremely difficult."[12] On June 28, 1976, the *Wall Street Journal* reported that the plant which "was supposed to open in 1974 isn't likely to get started until 1978, or even much later."

As a result of the delays in developing a reprocessing capability, and the continuing bottleneck that is anticipated in this area, a backlog of spent fuel is anticipated even with an optimistic schedule of operation of the proposed facilities. This is evident from Figure 2.8. The magnitude of the projected unbalance between reprocessing capacity and fuel requiring reprocessing is vividly illustrated in Figure 2.9. A severe shortfall of capacity will result, beginning with a 700 MT uranium processing deficiency in 1983 and growing thereafter.

D. E. Saire of ERDA states:

FIGURE 2.8

LWR Spent Fuel Reprocessing Supply and Demand (Cumulative Basis)

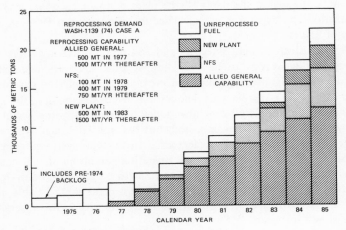

Source: U.S. AEC, WASH 1174-74, op. cit., p. 63.

FIGURE 2.9

Comparison of Industry Reprocessing vs. Domestic Uranium Returns

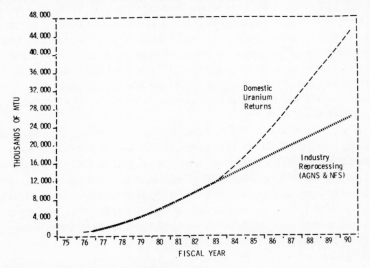

Source: U.S. ERDA, CONF-750209, op. cit., p. 71.

In order for the reprocessing industry to support the projected returns, three additional reprocessing facilities, 1,100 MTU/yr. (or two AGNS Barnwell capacity plants) are required with the first new capacity available by FY 1983. The second and third facilities would apparently be required in 18-month to 24-month sequence after the first new capacity comes onstream. In other words, new reprocessing capacity, beyond AGNS and NFS, will probably be required in FY 1983, 1985, and 1987 to support the projected domestic uranium returns shown in Figure [2.10]. . . . It should be noted that the current estimate of lead time to bring a new reprocessing plant onstream is between 8 to 10 years.[13]

FIGURE 2.10

Industry Processing Plans

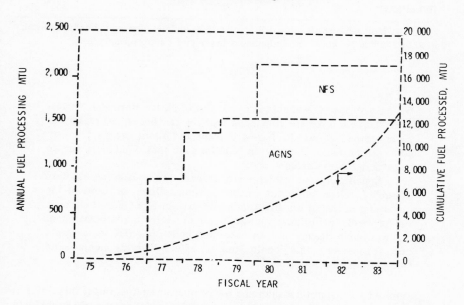

The total number of reprocessing plants and the estimated capital cost in 1975 dollars are shown in Table 2.6. The prospects at this time for additional plants being built are almost nil. "The fact that there is a lack of reprocessing facility is not accidental because reprocessing is 'marginal' economically."[14] Beyond the economic aspects, there is strong environmental opposition to reprocessing plants, which can hinder the timely building of further plants necessary to the proper functioning of the nuclear fuel cycle. Even such persons as Jimmy Carter, ex-governor of Georgia, and trained as a nuclear engineer, have expressed strong reservations. In a letter to the AEC, he wrote:

TABLE 2.6

Reprocessing Capacity and Capital Requirements

	Low Case	Moderate/ Low	Moderate/ High	High
MT capacity needed by year 2000	10,700	13,400	16.800	21,600
Less Barnwell & NFS plants	(2,250)	(2,250)	(2,250)	(2,250)
Capacity to be added	8,450	11,150	14,550	19,350
No. of 1,100 MT plants needed	7.68	10.14	13.23	17.59
1975 dollars (in billions)*	9.2	12.2	15.9	21.1

*The Barnwell plant is estimated to cost $500 million when completed. For escalation to $1.2 billion per plant, see methodology in Chapter 7 and Appendix D.

We have reviewed available material regarding the Barnwell Nuclear Fuel Plant. . . . We are convinced that Georgia has strong reasons to be concerned about the Barnwell plant and about the tendency to concentrate major nuclear installations of that kind adjacent to Georgia in South Carolina.

Specifically, we are concerned about the continuing build-up of such facilities bordering on the Savannah River, in terms of the increasing potential for direct radiation exposure of Georgia citizens, the probable continuing build-up of radioactive contamination of the Savannah River, and the steadily increasing risk of a major nuclear accident that could greatly affect Georgia people and Georgia's natural resources.[15]

Jimmy Carter is not considered a foe of nuclear energy, but if this action is symptomatic of future attitudes, this phase of the nuclear fuel cycle could be in serious trouble. Last year, *Nuclear News* wrote, "Since the [GE Midwest Fuel Recovery Plant] was the only fuel reprocessing plant scheduled for operation during the next several years, there is no alternative place for GE's customers to go, leading to a fast-approach glut of irradiated fuel. Nor will the utilities be able to contract for enrichment services until their fuel is recovered. An AEC spokesman said the AEC, GE, and Nuclear Fuel Services may be able to store some of it for a while, but in general, the situation is 'a catastrophe.'"[16] This shortfall in reprocessing will lead to significant problems regarding the storage of spent fuel, which in turn, can lead to shutdowns of nuclear plants.*

THE PROBLEMS OF SPENT FUEL STORAGE

When fuel is removed from a nuclear reactor, it is stored in a pool at the reactor site to permit short half-life radionuclides to decay, which takes about 150 days. These pools, however, are not designed as permanent storage sites for spent fuel, and their capacity is limited. An extract from WASH 1174 summarizes the problems of spent fuel storage:

> Delays in start up of reprocessing plants, and the possibility that future capacities of the plants will not keep pace with demand, suggest the possibility that a serious shortage in spent fuel storage capacity could develop. In addition to the spent fuel storage basins located at the reprocessing plants, some spent fuel storage capacity is provided at each of the power reactor sites in order that spent fuel may be discharged and allowed to cool before being shipped to reprocessors. The reactor basins are also designed with a second possible use in mind—that of providing space for removal of an entire core—if it should become desirable in order to expedite emergency maintenance. The reactor basins, therefore, usually have capacity to hold fuel assemblies equivalent to one full fuel core plus one or two reload batches; however, there are some situations in which two reactors share a single basin and the space allotted to each is equal to less than a full fuel core.
>
> Past, and possible future, delays in completion of reprocessing plants raise the possibility that a serious shortage in spent fuel storage capacity could develop in the near future. It appears likely that about five reactor site storage basins will be completely filled by the end of 1976 unless steps can be taken to transfer some of their fuel to other locations. Similarly, a number of additional reactor basins

*On July 15, 1976 a news article in the *Wall Street Journal* stated: "Nuclear Fuel Services, Inc., the only company ever to operate a commercial nuclear fuel reprocessing plant in the U.S. is moving toward a decision to end the venture. . . . An NFS pullback from the West Valley project, which a company spokesman said is 'under serious consideration,' would underscore the nuclear industry's continuing inability to deliver one of the economic benefits long claimed for atomic power plants. . . . Nuclear Fuel Services' move to serious consideration of a total pullout at West Valley was triggered by an April NRC decision to impose more stringent guidelines to protect the plant from earthquake-type disturbances. The ruling was in connection with the pending NFS application for the construction permit. . . . 'The seismic ruling was a serious enough blow to cause us to take a long hard look at whether we can ever make a go of the West Valley operations,' an NFS spokesman said."

will become completely filled in 1977 and subsequent years. The storage capacities presently planned for the reprocessing plants (approximately 360 metric tons at the AGNS plant and 400 metric tons at the NFS plant) will not be sufficient to prevent a serious shortage capacity, if the plants themselves fall significantly behind in startup schedules and levels of operation.[17]

At this time, the GE plant at Morris, Illinois is being used to store spent fuel, but the space is rapidly being exhausted. It is possible that GE may increase the capacity of the storage basins from 90 to 600 MT, but this is uncertain.

A. E. Schubert, President of Allied-General, the Builder of the Barnwell Plant, "suggested that about 40 percent of U.S. nuclear generating capacity could be shut down over the next five years if the government fails to expedite decisions on processing, handling, and storing nuclear wastes."[18] Earlier, the Bureau of National affairs reported, "If the startup of Allied [General's] facility is delayed until 1979 and that of NFS until 1980, the buildup of unreprocessed fuel would total about 5,800 metric tons of uranium by 1982."[19]

If there are inadequate facilities for storing spent fuel, negative occurrences will naturally ensue. Plants will be forced to close and there will be a shortfall in the quantity of electric energy available, resulting in potential brownouts, blackouts, and hindrance of economic expansion. Utility rates will rise significantly for consumers in those areas where plants are closed down. This is because the affected utilities will be forced to buy electricity off the grid at various negotiated rates, many of which will be higher than the cost of generating electricity from their own plants. And secondly, the consumer will still have to pay for the capital costs (i.e., interest and amortization) of the closed plant, which will remain in the rate base.

The alternative of not maintaining the closed plant in the rate base can only result in economic failure for the utility, which is inconsistent with the concept of virtually all federal and state utility regulation. Thus, the economic costs of this bottleneck can be very high.

RADIOACTIVE WASTE MANAGEMENT

The nuclear fuel cycle produces both high-level and low-level radioactive wastes. The high level wastes are the ones of most concern, and the largest percentage of them is produced as a by-product of reprocessing the highly irradiated nuclear reactor fuels. According to the Atomic Energy Commission, "These wastes have such a high radioactive content of long-lived isotopes [24,000-year half-life] that they require long-term storage in isolation and under essentially perpetual surveillance at the storage sites. Before storage, these wastes will be processed into inert, immobile, solid material which is nonexplosive, noncombustible, and cannot turn to gaseous form and become airborne."[20]

The AEC environmental survey of the uranium fuel cycle states:

Appendix F, 10 CFR 50, requires that the inventory of high-level liquid waste at a fuel reprocessing plant must be limited to that generated in the prior five years. The liquid waste must be converted to dry solids and transferred to a Federal repository no later than ten years following separation from irradiated fuels.

It was planned to construct a Federal repository in a salt mine for long-term geological storage of solid high-level wastes by the mid 1970's. However, subsequent events have deferred the site selection and construction of such a facility.

The present alternative plan is to design and construct a Retrievable Surface Storage Facility capable of safely storing high-level radioactive waste containers for 100 years or until such time as a permanent Federal repository can be established. . . . Although a repository for long-term geologic storage of high-level waste is intended to isolate the waste from man and the biosphere, the Federal government will have the obligation to maintain control over the site in perpetuity.

[Storage of the radioactive wastes would be simple except] for the fact that radioactive decay releases significant quantities of heat for significant time periods. . . . Should adequate provisions not be made to remove this heat and transfer it to the natural heat sink of the atmosphere, the waste and the canister would melt.[21]

This aspect of the nuclear fuel cycle is difficult to evaluate strictly in dollar terms. At this particular time, ideas for a Retrievable Surface Storage Facility (RSSF) are still being evaluated and there are no plans to begin construction until 1979 or later.

The radioactive waste management phase is one more step in the nuclear fuel cycle that has not been fully developed, although it should not be impossible to achieve in some way, shape, or form. However, previous experience in seeking a permanent disposal site was unsuccessful because of public opposition that developed when an abandoned salt mine near Lyons, Kansas was selected by the AEC.

SUMMARY

Three stages of the nuclear fuel cycle present no economic problems. These include conversion of ore concentrate to UF_6, conversion of enriched uranium to fuel material, and fabrication of fuel elements. However, all other steps entail one or more of the following problems: capital risks, resource shortages, schedule slippages, and/or technological problems.

In the area of mining and milling, there is a potential long term shortage of uranium ore, and by the end of the early 1980s, an anticipated shortfall in capacity to mine the ore that is available and will be required.

Regarding enrichment, slippages in plant expansion by ERDA could result in fuel shortages even for those utilities that have firm contracts. If forecasts of nuclear power growth are correct, numerous reactors coming on line during the mid 1980s will not be able to obtain fuel. High economic risks plus enormous capital needs have deterred private corporations from building the enrichment facilities that will be needed subsequent to 1983. Because of the 10-year lead time required, only slowdowns and delays in reactor construction can prevent shortages of enriched fuel from occurring.

Reprocessing is an integral element of the nuclear fuel cycle, although to date it has never been accomplished on a commercial scale. Shortages in capacity are forecast for a dozen or more years. This can exacerbate the shortfall of natural uranium over both the short and long terms. Even more seriously, the current bottleneck is creating storage difficulties for spent fuel. If this problem is not solved, an increasing number of reactors may be forced to close each year.

Radioactive waste management that requires isolating nuclear waste materials from the biosphere for at least 24,000 years has not even reached the final design stage. This is not an economic problem, but rather an environmental one, which if not resolved could eventually lead to legislation adverse to the continued development and use of nuclear power.

CONCLUSION

The nuclear fuel cycle may be characterized as a complex technology with many functional problems still unresolved. In the past, this has not caused significant difficulties because nuclear power production expanded slowly and made only light demands upon the entire nuclear system. In the future, however, the ability of nuclear reactors to deliver power will be dependent upon many technologies that have never functioned before on a commercial scale, and also upon these activities meeting design specifications on schedule. Yet, even if all planned facilities perform as forecast, there will be capacity shortages in mining, enrichment, and reprocessing.

All of these factors combine to multiply the risks to the utility in failing to meet its mandate of providing adequate electric power; to the consumer who may not only be without power, but may have to pay for idle plants; and finally, to the investor. The successful functioning of the nuclear fuel cycle is contingent upon enormous capital investment in processes such as enrichment or reprocessing, which are marginally economic, have attendant high risks, and are significantly less attractive than other investments such as petrochemical plants.

Coupled with environmental opposition, it is questionable whether nuclear power can be an attractive investment.

NOTES

1. General Advisory Committee to the U.S. Atomic Energy Commission, 130th Meeting, November 7, 1974, Uranium Program, Division of Production, Division of Production and Materials Management, U.S. AEC.

2. Uranium Enrichment Conference, Oak Ridge, Tennessee, February 13, 1975, conducted by ERDA, CONF-750209, p. 81.

3. Ibid., p. 4.

4. Ibid., p. 24.

5. Ibid., p. 61.

6. Ibid., pp. 45 and 61.

7. AEC, *Nuclear Fuel Resources and Requirements*, WASH 1243 (April 1973): 46.

8. Charles Joslin, "Uranium Enrichment," *Barrons*, July 7, 1975, p. 11.

9. General Advisory Committee to the AEC, op. cit.

10. Bureau of National Affairs, Bulletin Number 91 (May 8, 1975): D-6.

11. *Science*, Volume 185 (August 30, 1974): 771.

12. AEC, *The Nuclear Industry*, WASH 1174-74 (1974): 62.

13. ERDA, CONF-750209, op. cit., p. 68.

14. Bureau of National Affairs, loc. cit.

15. Governor James Carter, Letter to Robert M. Lazo, Esq., Chairman, Atomic Safety and Licensing Board Panel, U.S. Atomic Energy Commission, dated January 6, 1975.

16. *Nuclear News*, August 1974, p. 64.

17. AEC, WASH 1174-74, loc. cit., p. 63.

18. Bureau of National Affairs, op. cit., p. D-5.

19. Bureau of National Affairs, Bulletin No. 72 (December 26, 1974): A-6.

20. AEC, WASH 1174-74, op. cit., p. 64.

21. AEC, *Environmental Survey of the Uranium Fuel Cycle*, WASH 1248 (April 1974): G-6, 7.

3

COAL AVAILABILITY
AND COST

Just as with nuclear fuel, the amount of reserves, the availability of these reserves, and production cost will determine the competitiveness of coal as a primary energy source. Along with these basic factors, the impact of some secondary influences must also be considered; namely, the availability of capital and equipment, the adequacy of transport delivery systems, and the potential environmental constraints on strip mining and air pollution.

ADEQUACY OF COAL RESERVES

Coal is the most abundant nonrenewable energy resource in the United States. The magnitude of these resources can be seen in Table 3.1. These gross tonnage figures do not represent a homogeneous commodity, but rather one with considerable variations regarding method of mining, sulfur content, heat value, geographic location, and environmental impact. All of these factors affect the true availability of the demonstrated reserves.

The first major classification of coal is by method of mining—underground or surface. Of the 398 billion tons of demonstrated bituminous and sub-bituminous coal reserves available for electrical generation, approximately 27 percent would be surface mined and 73 percent underground mined.[1]

Two-thirds of the coal resources are considered to be low-sulfur with a sulfur content of 1 percent or less;[2] almost all of the low sulfur reserves are in the western states while only 6 percent is found in the traditional coal producing areas of Alabama, Illinois, Indiana, Kentucky, Ohio, Pennsylvania, Virginia, and West Virginia.[3] Although western low-sulfur coal is normally thought of as being surface mined, almost 63 percent or 118.2 billion tons is classified as deep-mined

TABLE 3.1

Total Coal Resources

Demonstrated reserves (economically attractive with today's technology based on today's prices)	434 billion tons
Other identified reserves	1,166
Total identified reserves	1,600
Additional postulated but unidentified resources	1,600
Total potential resources	3,200 billion tons

Note: All tonnage figures in this chapter are short tons of 2,000 lbs.
Source: U.S. Federal Energy Administration, *Project Independence Blueprint, Final Task Force Report: Coal*, November 1974, p. 4, citing U.S. Bureau of Mines, Division of Fossil Fuels, Mineral Supply, *Demonstrated Coal Reserve Base of the U.S. on January 1, 1974*, June, 1974.

coal.[4] Surface mineral coal not only is cheaper to obtain, but has recovery rates up to 90 percent, while underground coal mined by the room and pillar method has only a 50 percent recovery rate. However, with new investment and introduction of the longwall method used in Europe, even the deep-mined recovery rates can be increased to 85 percent

There are four major types of coal mined in the United States—bituminous, subbituminous, anthracite, and lignite—with bituminous and subbituminous being the predominant ones used for electrical generation. In addition to variations in sulfur, coal from different regions varies in its combustion, conversion, and coking characteristics as well as ash level and heat value. By far the most important factor of these is that of heat value, which can range from a high of 15,350 Btu/lb. for West Virginia bituminous to a low of 8,890 Btu/lb. for Washington state subbituminous and 7,000 Btu/lb. for Texas lignite. Eastern and midwestern coal, which usually has a sulfur content in excess of 2 percent, has the highest heat values, tending to average around 13,500 Btu/lb., while western coal, which is virtually all low-sulfur, has lower heat values averaging around 12,250 Btu/lb.[5] A higher heat rate not only means that less coal need be burned for each kwh of electricity generated, but it also affects the sulfur limits set by the Clean Air Act. For example, without scrubbers or other sulfur cleaning equipment, coal with a heat value of 12,000 Btu/lb. is limited to a sulfur content of 0.7 percent by weight, but 8,500-Btu/lb. coal has an even lower limit of 0.5 percent sulfur.

BARRIERS TO COAL USAGE

Two types of obstacles can slow the rapid expansion of coal usage in the immediate future:

1. problems related to sulfur pollutants
2. constraints on strip mining

Although western coal is low in sulfur, on a weight basis, much may not meet new source performance standards. Nonetheless, because engineering progress rather than a scientific breakthrough is required, it would not seem unreasonable to expect the problem of sulfur emissions to be resolved within the next few years. This could be accomplished through improved flue gas desulfurization systems, through conversion of coal to cleaner fuel forms either at the point of production or at the point of consumption, or through some other approach.

Assuming that the performance of the gas desulfurization systems for removing excess sulfur from coal burning exhaust gases proves to be satisfactory, adequate supplies of these systems should be available. The Environmental Protection Agency, in a recent survey of vendors, found that 131,000 MWe of capacity could be equipped by 1981 with existing production facilities, and 197,700 MWe with expanded facilities. This latter figure is greater than the total generating capacity that the EPA expects will have to meet sulfur control requirements by that date.[6]

Strip mining, especially in the West, offers the potential not only of low-sulfur coal, but of reserves easily accessible at the lowest cost.

Discussions with utility executives in July 1975 indicate that the veto by President Ford of the compromise strip mining bill has created an air of uncertainty regarding surface mining. The veto of this bill may have won the battle for coal companies but effectively lost the war. From conversations held in March 1975, coal companies stated that were the bill to pass, they were prepared to comply with the restrictions and to proceed with the opening of new mines.* Although the bill was not as strong as environmentalists had wished, it probably would have defused the issue, but now an alliance of environmental groups called the Strip Mining Coalition is seeking a total ban on all surface mining.

While a total ban, which was part of an overall federal policy with a coordinated conservation plan, would probably not hinder national economic growth, a bill that solely seeks to end strip mining conceivably would not pass Congress because of the theoretical argument that unemployment would result.

---

*See Appendix C for a list of conversations.

Regardless of the odds associated with the passage of any strip mining legislation, sufficient uncertainty has been created which is manifesting itself as a slowdown of investment.

Considering that there are constraints, either technological or environmental, on the mining and use of coal, a conservative approach would exclude up to half the coal reserve base from a discussion of available usable resources. With this assumption, it would appear feasible to project 199 billion tons as the quantity of demonstrated reserves available for electrical generation. This figure then becomes the basis for comparison with projected coal requirements.

COAL DEMAND

Using the ERDA projections (see Appendix A) for fossil fuel plants, and assuming that by 1980 all fossil fuel plants, except internal combustion or gas turbines, have been converted to coal burners, the annual usage of coal can be derived as shown in Table 3.2.

An interesting trend in these projections is that by 2000, the low growth case would require almost twice as much coal as the high. This apparently results from a greater coal commitment with the low case where the capacity factor is shown as 46.6 percent, indicating predominantly base-loaded plants. This contrasts with the high case where the capacity factor drops to 27.6 percent, an

TABLE 3.2

Annual Coal Demand for Electrical Generation
(Millions of short tons—average heat rate: 10,100 Btu/lb.)

Year	Low Case	Moderate/ Low	Moderate/ High	High
1980	915.7	931.5	929.0	936.1
1985	1,082.0	1,023.6	999.1	981.2
1990	1,117.0	1,074.7	1,001.2	932.0
1995	1,196.8	1,139.0	972.4	835.0
2000	1,378.7	1,272.8	973.8	737.8

Note: Coal Usage = (Btu $\times$ 10^{12})/(20.4 Btu $\times$ 10^6/ton). The heat rate of 10,200 Btu/lb is on the low side and highly conservative, thus probably overstating coal usage somewhat.

Source: Compiled by the authors.

TABLE 3.3

Maximum Annual Coal Demand Assuming No Nuclear Expansion After 1985
(Millions of short tons—average heat rate: 10,200 Btu/lb.)

Year	Low Case	Moderate/ Low	Moderate/ High	High
1980	915.7	931.5	929.0	936.1
1985	1,082.0	1,023.6	999.1	981.2
1990	1,478.0	1,504.9	1,509.7	1,613.3
1995	1,989.8	2,132.7	2,188.6	2,461.2
2000	2,636.3	2,921.5	3,125.3	3,621.4

Source: Compiled by the authors.

indication that fossil plants would be almost exclusively used for intermediate and peak-loading purposes.

Because of the potential shortages of uranium forecast in Chapter 1, a "maximum annual coal demand" table was derived in order to "test" the adequacy of coal reserves as a reliable future fuel source. The assumption was made that subsequent to 1985, no additional nuclear capacity would come on line and that all projected nuclear plants would be constructed as coal burners instead. Tables 3.2 and 3.3 now become the basis for projecting the cumulative coal demand for electrical generation to the year 2000 as shown in Table 3.4.

DEMAND VS. SUPPLY

With a maximum usage based on the "high case," and assuming that no nuclear plants come on line subsequent to 1985, cumulative demand for coal to the year 2000 is 41.0 billion tons. This compares with "demonstrated reserves" of electrical generating coal of 199 billion tons, predicated on a 50 percent recovery factor.

Based on growth rates of 1.5 percent and 3 percent from the year 2000 on, Table 3.5 gives the years in which the different levels of resources would be depleted.

TABLE 3.4

Cumulative Coal Demand for Electrical Generation
Based on ERDA Projections
(Millions of short tons—average heat rate: 10,200 Btu/lb.)

Year	Low	Moderate/ Low	Moderate/ High	High
As Shown by ERDA				
1973*	400	400	400	400
1980	7,052	6,142	6,095	6,142
1985	11,962	10,990	10,881	10,917
1990	17,481	16,211	15,978	15,724
1995	23,236	21,721	20,984	20,095
2000	29,590	27,707	25,944	24,057
No Nuclear Expansion After 1985				
1973*	400	400	400	400
1980	7,052	6,142	6,095	6,142
1985	11,962	10,990	10,881	10,917
1990	18,184	16,995	16,861	16,970
1995	26,515	25,649	25,718	26,562
2000	37,732	37,915	38,559	41,002

*Actual.
Source: Compiled by the authors.

TABLE 3.5

Year of Coal Resource Depletion (50 percent Recovery Rate)

	Growth Rates After Year 2000	
	1.5 Percent	3 Percent
Demonstrated Reserves	2034	2028
Identified Resources	2095	2066
Postulated Resources	2135	2089

Source: Compiled by the authors.

Assuming that:

1. No new coal discoveries are made,
2. Technological improvement does not bring an increase in the recovery rate,
3. No other energy sources, such as solar, are developed, and,
4. Coal is virtually the only primary source of stational electrical generators,

the U.S. coal resources will last well into the twenty-first or twenty-second centuries.

Since these assumptions are so absurdly conservative, the only conclusion that can really be drawn is that the United States has more than an abundance of coal to meet its needs for a great many years.

COAL PRICES

While newspaper headlines since 1973 have highlighted exorbitant coal prices paid by some utilities, these were solely "spot" prices for immediate or near-term delivery paid under adverse conditions. Most purchases now negotiated by utilities are long-term 20- to 30-year contracts predicated upon a base price at the mine. Added to the base price are production escalators that vary with changes in the costs of mining, such as labor, equipment, or new environmental requirements.

A March 1975 survey of coal producers in the western states yielded average mine head quotes of $8/ton for surface-mined coal and $15/ton for deep-mined. Illinois coal was quoted at $15/ton for surface-mined and $20/ton for deep-mined, the latter figure reflecting the increased costs associated with new safety rules. (See Appendix C for tabulation of survey results.) This converts to electrical generating fuel costs as shown in Table 3.6. (Mileage costs are given for unit train shipments.)

To calculate the delivered cost of fuel for a specific location, multiply the distance between mine and utility by the mileage cost in the "mills/kwh" column and add to the cost of coal.

For example: Hanna, Wyoming to Des Moines, Iowa = 800 miles
      800 X .002262   = 1.81 mills/kwh
      Cost of strip-mined coal   = 3.62
         Total fuel costs   = 5.43 mills/kwh
Marion, Illinois to Des Moines, Iowa = 470 miles
      470 X .00891   = 0.89 mills/kwh
      Cost of strip-mined coal   = 5.67
         Total fuel costs   = 6.56 mills/kwh

TABLE 3.6

Typical Coal Cost

	$/ton	¢/million Btu	Mills[a]/kwh
Wyoming			
Strip	8.00	38.10	3.62
Deep	15.00	71.43	6.79
Mileage Cost[b] (per			
ton-mile)	0.005	0.02381	0.002262
Illinois			
Strip	15.00	59.71	5.67
Deep	20.00	79.62	7.56
Mileage Cost (per			
ton-mile)	0.005	0.0199	0.001891

Heat Values: Wyoming = 10,500 Btu/lb.

                       Illinois = 12,560 Btu/lb.

Heat Rate = 9,500 Btu/kwh

Mileage rate is for unit trains with 100-ton cars, 100-150 cars/train.

---

[a]One mill = 0.001 dollar, or 0.1 cent.

[b]This is based on a specific tariff in effect in March 1975 for hauling Wyoming coal to Sioux City, Iowa in freight cars owned by a utility. Every unit train run will have its own unique tariff, which will vary widely with little correlation between different railroads and locations even though mileages may be nearly the same. The large discrepancies between rates that should be similar is one of the idiosyncrasies of Interstate Commerce Commission (ICC) regulation.

Note: To convert $/ton to cents/million Btu:

$$(\$/\text{ton} \times 100¢/\$ \times 10^6) / (\text{Btu/lb.} \times 2{,}000 \text{ lb./ton}) = ¢/\text{million Btu}$$

To convert ¢/million Btu to mills/kwh:

$$\frac{¢ \times 10 \text{ mills}/¢}{10^6 \text{ Btu}} \times \frac{\text{Btu}}{\text{kwh}} = \frac{¢ \times \text{Btu in thousands}}{100} = \text{mills/kwh}$$

The cost effect of a total ban on strip mining would, of course, be to limit the choice of coal to the latter categories of deep-mined resources.

AVAILABILITY OF CAPITAL AND EQUIPMENT

The costs of opening a new mine are not great, especially when compared with the cost of a single 1,000 MWe generating plant. To support a plant of this size, a mine would require an investment of $41.0 million to $82.2 million.[7] Because utilities sign 20- to 30-year contracts and the risky exploration associated with petroleum can be avoided, financial institutions find coal mines to be attractive investments, and adequate capital is generally available. Key materials required for mine openings would include items such as roof bolts, electric trailing cable, conveyor belting, and dragline equipment. Because this latter item has been in particularly short supply, Bucyrus-Erie, one of the largest manufacturers, was contacted. Based on current production capacity, this company as of early 1975 was back-ordered to 1979. However, according to W. G. Piper, manager of large machine sales, $100 million was being invested in a new plant that would quadruple the output of dragline equipment, thus reducing the backlog within the foreseeable future. Government energy officials confirmed that similar expansion was occurring in other companies.

THE TRANSPORT OF COAL

There are four methods for transporting coal, with railroad being by far the most prevalent.

TABLE 3.7

Coal Transportation in 1972

Method of Transport	Tons (Millions)	Percent of Total
Railroad	379	66
Water	70	12
Truck	70	12
Coal slurry pipeline	3	—
Used near mine*	62	10

*Used at mine-mouth steam-electric plants.
Source: FEA, Coal, p. 14 (see note 1, p. 59).

While coal slurry pipelines could conceivably be the most economical method, there is presently no national legislation that grants the right of eminent domain to slurry pipelines, and railroads generally will not grant construction permission for use of their right-of-way, since a pipeline would deprive railroads of coal-hauling revenue. Another factor hindering slurry shipment is the water requirement of 2-3,000 acre-feet per year for a 1,000-MWe plant. In view of the competition for water by agriculture and ranching, especially in the western states, it seems that this can become a practical barrier to any sizable expansion of the slurry approach.

Truck transport is hardly cost competitive under most circumstances. In fact, from 1965 to 1972 it declined not only percentagewise (15 to 12 percent) but also in absolute tonnage (76 million tons to 70 million tons).

Water conveyance can be especially attractive as an important mode of transport. Substantial expansion, however, would require federal assistance to obviate marine bottlenecks. This would encompass federal support for financing the construction of shoreside facilities and new cargo handling equipment as well as additional barges.

Rail transport is, and will continue to be, the primary carrier mode for coal. In order to accommodate the increased demand for coal, more hopper cars will be required, additional spur lines will have to be laid, and trunk and branch lines will have to be upgraded. According to the Association of American Railroads, new spur rail lines can be built for an average of $200,000 per mile. In the overall cost structure of a 1,000 MWe plant, this is not a significant cost. Nor is lead time a problem. While a plant takes six years to construct, and a mine two-and-a-half to five years to open, a new railroad line can be laid in a year.[8]

Much more critical in the overall scheme of getting coal from the mine to the plant is the general condition of roadbeds around the country. For profitable railroads such as the Union Pacific, roadbeds are well maintained and unit coal trains can travel at moderate to high speeds. However, when a unit train, for example, crosses the Missouri River from Nebraska to Iowa and then must travel on the tracks of the bankrupt Rock Island Line, the heavy unit trains must slow down to 20 to 30 mile per hour speeds, considerably decreasing their productivity. Similar conditions exist throughout the Northeast. While this situation poses an immediate threat to efficiency, it serves to point up the need for an overall federal energy program that would include railroad upgrading as an integral part of the plan.

A projection of doubled coal usage often carries with it a prediction of the need to double the number of hopper cars from the current 500,000 level. The fallacy in this forecast is that it assumes the continuation of old utilization patterns where coal travels as part of a mixed cargo freight shipment, with individual cars averaging only 15 revenue trips annually.[9] Performing this type of straight-line extrapolation is obviously absurd since it would assume that a utility would normally choose to pay 2.8 cents/ton-mile[10] instead of the unit train* cost of about 0.54 to 0.80 cents.[11]

Assuming that the maximum growth in fossil fuel plants occurs under ERDA projections between 1975 and 1985, the equivalent of 139 additional 1,000 MWe coal plants will have come on stream. In addition, if all other fossil plants are assumed to convert to coal, this would add a requirement for another 150 equivalent plants, or a total of 289 new 1,000 MWe plants to be serviced in 1985. Furthermore, if it is also assumed that there is no improvement in the utilization patterns of existing cars and that all additional coal is carried by rail, then for the ten-year period, 115,600 new hopper cars would be required.† Since 16,174 new hopper cars were installed during the peak year of 1971,[12] the 11,560 annual requirement should pose no manufacturing problem. With a $30,000 to $40,000 price per car, the $16 million requirement per plant is only a small percentage of the overall cost of a facility.‡

---

*A unit train is a complete train of up to 100 or more cars that carries only one product (in this case coal) following only one route between the producer and the consumer. A unit train would load at the coal mine (slowing down to a few miles per hour but not stopping), travel directly to the utility plant where it unloads (again slowing down but not stopping), and then returning to the mine for another load. The train always operates as a complete unit, neither uncoupling nor stopping except for maintenance or crew changes. As a result of this concept, car utilization reaches levels of 200 to 300 trips per year.

†*Method of Computation*

Assumptions:

    1,000 MWe plant (10^6 kw plant)

    Distance from mine to plant = 1,000 miles

Time for round trip: (2,000 miles)/40 mph   = 50 hours

                                            +   5 hours loading

                                            +   5 hours unloading

                                                60 hours total

Plant coal requirement:

    (10^6 kwh × 24 hours × 9,500 Btu/kwh) / (20.4 × 10^6 Btu/ton of coal)

    = 11,176 tons per day

    (11,176 tons/day)/(24 hours/day) × 60 hours = 27,940 tons/60 hours

One unit train   = 100 cars × 100 tons/car

                                         = 10,000 tons/train/60 hours

Plant requirements = (27,940 tons/60 hours) / (10,000 tons/train/60 hours) = 2.79 trains ≈
                                                              3 trains

3 trains × 100 cars/train   = 300 cars

Reserve for maintenance, etc.   = 100 cars

Total plant requirements   = 400 cars/plant × 289 plants = 115,600 cars

‡Most utilities now buy their own cars.

SUMMARY

With the highest electric usage growth rate projected by ERDA (i.e., "high case") to the year 2000, and a 3 percent rate thereafter, demonstrated reserves of coal will last until 2028, and all postulated additional resources until 2089. This assumes no improvements are made in the technology of mining coal.

In the immediate future, the major barrier to increased coal usage is the problem of controlling sulfur emissions. Utilities are reluctant to install stack gas scrubbers; they claim that the engineering is not perfected at this time and feel that a more effective technology such as "pre-washing" of coal will be available within the next few years. On a long-term basis, there are no technological barriers to the usage of coal as the major stationary energy source of the United States. However, a national commitment is required to overcome any institutional and financial barriers that may arise.

CONCLUSION

As a long-term strategy for supporting the bulk of the stationary energy needs of the United States, it can be seen that there are no technological or economical impediments to the utilization of coal. The most immediate need is to improve processes for the removal of sulfur or coal exhaust gases. Strip mining presents no technological barrier, but rather a political one based upon uncertainty of regulation. Even if a total ban on strip mining were enacted, however, there would be sufficient coal reserves to meet the needs of electric utilities. In addition, there must be a definite national energy policy in order to avoid any potential bottlenecks that might occur within the coal energy system.

NOTES

1. U.S. Federal Energy Administration (FEA), *Project Independence Blueprint, Final Task Force Report: Coal*, November 1974, p. 105, citing U.S. Bureau of Mines, Division of Fossil Fuels, Mineral Supply, *Demonstrated Coal Reserve Base of the U.S. on January 1, 1974*, June 1974.

2. Ibid., p. 5.

3. U.S. Environmental Protection Agency (EPA), *Evaluation of Sulfur Dioxide Emission Control Options for Iowa Power Boilers*, EPA report 650 12-74-127, (December 1974): 6.

4. FEA, op. cit., p. 105.

5. Ibid., Table 2, pp. 106-12.

6. EPA, *Report to Congress on Control of Sulfur Oxides*, EPA-450 11-75-001 (February 1975): 2, 30.

7. FEA, op. cit., pp. 21-22. The figures given by the FEA for early 1974 were $31.5 million to $63.2 million, which were then escalated by 30 percent inflation.

8. Interview with Norm Linse, Market Manager of Energy Resources, Union Pacific Railroad, March 11, 1975.

9. *Railway Management Review*, Volume 73, No. 3, p. A-96.

10. EPA, EPA-6501 (February 2, 1974); EPA 650 12-74-127, op. cit., p. 37, Iowa utility transportation costs for 1974.

11. Union Pacific Railroad, published rate schedules.

12. Interview with Kenneth H. Hurdle, Chief Statistician, Association of American Railroads, August 1975.

4

HOW THE PRICE
OF ELECTRICAL
POWER IS DETERMINED

In order to comprehend fully the economics of power generation, an understanding of the approach to pricing is helpful. Since the distribution of centrally generated electrical power within a given geographic area is usually a monopoly, the rates are controlled either by a state or local agency, or under certain conditions by the Federal Power Commission.

THE THEORY OF RATE DETERMINATION

Over the years, a semi-uniform methodology has evolved for calculating the price of power, and the first question involved in rate making is: "What is the utility's total 'cost of service'?" To put it another way, "How much in total revenue should the utility be authorized to collect for its electricity?" This is often referred to as the "revenue requirement" or "cost of service," which is defined as the total of 1) proper operating expenses, 2) depreciation expense, 3) taxes, and 4) a reasonable return on the net valuation of the property devoted to public service.

The total net value of the company's tangible and intangible capital is called the "rate base" for rate-making purposes. The rate base is composed principally of the net (or depreciated) valuation of the utility's tangible property, which comprises the plant and equipment devoted to serving the public; also, an allowance for working capital and, depending upon the circumstances, amounts for other prescribed intangibles.* There are two methods for

---

*In essense, if the utility is *not* engaged in non-utility enterprises such as operating a busline or owning coal mines, the "rate base" will correspond to the asset side of the balance sheet.

valuing a company's tangible property: 1) original cost less depreciation, and 2) "fair value" (varying repriced investment methods).

In most cases, the original cost less depreciation is chosen as the valuation, and on this "rate base" the utility will earn a "rate of return," which is an amount over and above operating expenses, depreciation expense, and taxes, expressed as a percentage of the rate base. Within the rate base will be various classes of capital: debt, preferred stock, and common equity, each with its own average cost of capital. A typical cost of capital computation may look like this:

TABLE 4.1

Example of Cost of Capital Computation

Type of Capital	Amount of Capital	Percent of Total Capital	Avg. Cost of Each Type of Capital	Composite Cost
	(millions of dollars)		(percent)	(percent)
Debt	180	50	7.5	3.75
Preferred stock	36	10	8.5	0.85
Common stock, retained earnings, and surplus	144	40	12.75	5.10
				9.70

Source: Compiled by the authors.

From a rate-setting standpoint, the important figure is not the composite cost of capital, but rather the "return on equity" (ROE), which is the amount set by the regulatory agency. The cost of debt and the dividends on preferred stock are easy to ascertain, but the "fair return on equity" is set by a regulatory agency. In this particular example, the utility would be entitled to a "return on equity" or "profit" of $18.36 million, which is 12.75 percent of the $144 million equity value of the company.* This $18.36 million is computed after the payment of federal and state income taxes.

---

*The "return on equity" of 12.75 percent is the current percentage used in Iowa, which, compared with other states, is on the low side. The range nationally is 12 to 16 percent, and most utilities now argue that a 15 percent ROE is necessary (which may be true) in order to be able to finance future construction.

When a plant is being constructed, the relevant capital figure generally becomes the composite cost of capital, which is allocated and added to "brick and mortar" costs. For example, if a company were constructing $50 million of new plant in a fiscal year, the final amount that went into the rate base would be $54.85 million, which represents $50 million for "brick and mortar" and $5.85 million for capital costs, which is 9.7 percent of the $50 million spent on "brick and mortar." (In practice, the computation is made monthly as the funds are expended.) This total amount is added to the previous construction costs to obtain a total cost of new plant, but the dollar value is normally not entered into the rate base and charged to the consumer until the plant enters service.* After the overall "cost of service" has been established, the utility then devises a rate schedule that will yield the revenues required. Pricing is usually done in mills per kilowatt-hour.

The economic models in this study will not account for depreciation, which is a bookkeeping non-cash expense. Rather, the analysis will be from a cash flow standpoint which is a truer economic measurement.

GENERATION COST VS. DELIVERED PRICE

The price that a utility customer pays for his or her power is the "delivered price" of that electricity. In Iowa in 1975, a typical delivered price for residential electric customers was 4¢/kwh. This delivered price of power reflects *all* the costs of the utility service, which includes not only the costs of generating power at the central plant, but also transmission lines, substations, billing, payroll, management, return on investment (ROI), and so on. The delivered price can be expressed:

Cost of generating power + all other costs = delivered price

In Iowa, the cost of generating power was typically less than 2¢/kwh in 1975, or less than half the delivered price of electricity to residential customers. Since it is only the cost of actually generating electricity that can be affected by decisions about what kind of power plant to build, the economic comparisons in this book are confined strictly to "generation costs" (except in Appendix F on

---

*In a number of states, however, such as Pennsylvania, "Construction Work in Progress" (CWIP) or "Allowance for Funds During Construction" (AFDC) are included in the rate base as incurred during construction. This approach is heavily favored by utility companies but is usually opposed by regulatory agencies, and is disadvantageous to consumers since they are paying a return on plant that is not delivering electricity. It is analogous to a landlord charging a future tenant rent during construction before the tenant has moved in.

solar power). For a central power generation system, typically this "generation cost" figure would be approximately doubled to obtain the cost of "delivered power."

FACTORS COMPRISING GENERATION COSTS

Three elements make up the costs of generating power: 1) capital costs, 2) fuel costs, and 3) operation and maintenance costs. "Capital costs," as mentioned earlier, include the "brick and mortar" costs as well as the "cost of money." This plant cost will then be amortized over a given number of years (typically 30 for a central station) with ROI computed. Using a $1 billion plant as an example, the computations would be:

Total cost of plant	= $1 billion
Rated output	= 1,000 MWe
Life of plant	= 30 years
ROI	= 15 percent

For 15 percent/30 years, the "capital recovery factor" (CRF) is 0.1523. (The "CRF" is a constant that takes into account the economic life of a capital facility or the period in which a company will seek to recover its investment plus a return on its investment.*) For a $1 billion plant, the ROI would be:

---

*Since 15 percent is a common "return on investment" goal for most major corporations and is also the rate of return that utilities are claiming they require in order to finance growth, this 15 percent figure is used. A 15 percent capital recovery factor is also justified on the following basis: Assume that a typical utility has a balance sheet with 55 percent debt at an average interest rate of 7.5 percent and 45 percent equity, on which it is entitled to a 12.75 percent after-tax return on investment, which is equivalent to 25.5 percent pre-tax. The computation is:

.55 × .08 = .041
.45 × .255 = .115
$$\overline{.156 = \text{ROI}}$$

On this basis, 15 percent might even be low, because the average cost of debt is rapidly rising since utilities are increasingly selling large amounts of new, high-interest debentures. Because ROI implies an inherent profit, no "profit" figure, per se, is added in subsequent calculations.

$1 billion $\times$ 0.1523 = $152.3 million

This figure of $152.3 million will be the annual allocation of capital cost that can be charged consumers, and represents amortization of the facility plus a 15 percent ROI, which will be spread over the total units of electricity produced. A 1,000 MWe plant operating 65 percent of the time would result in a kilowatt-hour cost of 26.77 mills. This is calculated:

8,760 hours/yr $\times$ 65 percent capacity factor = 5,690 hours on line

1,000 MWe = 1,000,000 kw

5,690 hours $\times$ 1,000,000 kw = 5,690,000,000 kwh

($152,300,000) / (5,690,000,000) = $0.02677/kwh = 26.77 mills/kwh

This is the "capital cost" component of the retail price of a unit of electricity.

"Fuel costs" are the second major element of generating electricity, and have no fixed costs, but are variable depending on the production of power. The rate at which fuel is turned into electricity is fairly constant for any specific generating facility, and hence raw fuel costs can be translated directly into "fuel costs per kilowatt-hour" if the conversion rate, or "heat rate," of a power plant is known. This is expressed in Btu's per kwh, and fuel prices are usually expressed in cents/million Btu's ($\cent$/Btu). As an example:

Coal price	$\times$	Heat rate	$\times$	Conversion, cents to mills	=	mills/kwh
80$\cent$/MBtu		9,500 Btu/kwh		10		= 7.6 mills/kwh

"Operation and maintenance" (O & M) costs are minor compared with capital or fuel costs, and reflect such items as labor, maintenance, and the running of pollution control equipment. For a coal plant, this might be 2.0 mills/kwh. The "generating cost" of electricity will therefore be the sum of capital costs, fuel costs, and O & M costs. In our example this would be:

Capital cost	= 26.77 mills/kwh
Fuel cost	= 7.60
O & M cost	= 2.00
Total	= 36.37 mills/kwh

In order to reach an economic decision, all three costs must be included since low fuel and O & M costs can be outweighed by high capital costs, or vice versa. This will become readily apparent in subsequent chapters.

CONCLUSION

The key to having the lowest feasible rates is to insure that a utility is prudent in its investment decisions; that it doesn't make risky investments or overbuild; that sound business practice in all areas of management is followed, such as making a very determined effort to purchase fuel at the lowest available prices; and finally to insure that a utility is not playing games with its books and its earnings in order to obtain higher rates. Somewhere between the maximum rate a utility seeks and the minimum rate that could be charged in the short-term lies the middle ground that represents the best long-term value to the consumer.

The "capacity factor" is a measure of the reliability of a plant. It is a critical number because it directly affects the ultimate price of power. The cost of building a plant is a fixed amount that must be amortized over the number of kilowatt-hours produced; thus, the more electricity sold, the lower the allocation of fixed cost per kwh. With numerous questions being raised about nuclear plants because of capacity factors significantly lower than projected, an inquiry into the reliability of nuclear and coal plants is essential to a thorough economic analysis.

DEFINITION OF TERMS

"Availability" is a term often used to describe the performance of a plant. It is not a precise term since it states only that a plant was "available" for operation and that it produced some electricity. The formula for calculating annual availability is:

(Hours on line during a year) / (8,760 hours per year) $\times$ 100

In judging the efficiency of a plant, the availability factor is not a very accurate representation of performance since it merely measures the time during which a plant was operating. It would be similar to measuring the performance of an automobile designed to travel 55 miles per hour simply by recording the number of hours spent on the road even if it could not run at a speed greater than 20 miles per hour.

Capacity, on the other hand, measures the total designed performance of a plant against what it actually produced. Specifically, "capacity factor" is a

67

measure of the actual performance of a plant compared with its design power level. The formula for computing capacity factor is:*

$$\frac{\text{Actual kwh produced in a year}}{8,760 \text{ hours} \times \text{design power level of a plant (in MWe)}} \times 100$$

Theoretically, no plant can operate at a 100 percent level unless it is continually functioning at an output higher than its design level, since a shutdown for maintenance and repairs is mandatory at some time during the year. For a nuclear plant, refueling is usually a month-long job, thus reducing the maximum attainable capacity factor to 92 percent.[1]

THE EFFECT OF CAPACITY FACTORS ON ELECTRICITY COSTS

Since the cost of constructing a plant must be recovered by charging a portion of the plant to its customers each year, the plant will be amortized over the number of kilowatt-hours of electricity sold. Thus, the more energy produced, the lower the capital cost per unit. Assuming a fixed charge of 15 percent and capital costs of $640/kw for a nuclear plant and $525/kw for a coal-fired plant, the effects of different capacity levels upon the costs of power can be seen in Table 5.1.

While the low capital costs of nuclear plants in the past may have rendered them economical at capacity factors as low as 35 percent, it is clear that at today's construction prices, low capacity factors can heavily inflate electricity costs. At a capacity of 55 percent, the effective capital cost is 36 percent higher than that projected at the base level of 75 percent.

HISTORICAL RESULTS

While most industry, utility company, and AEC projections have used capacity factors of 70 to 80 percent for both nuclear and coal plants, the history of operations would indicate that these figures are overly optimistic. Byron Lee of Commonwealth Edison has said, "Although we would be the first to admit that we would like to see an 80 percent capacity factor on our base-load units,

---

*The definitions of "availability and "capacity factor" are those used by the Nuclear Regulatory Commission.

TABLE 5.1

Hypothetical Example of the Effects of Capacity Factors Upon Capital Costs

Capacity Factor (percent)	Cost as a Percent of Base	Nuclear (mills/kwh)	Coal (mills/kwh)
75	100	14.6	12.0
70	107	15.6	12.8
65	115	16.8	13.8
60	125	18.2	14.9
55	136	19.9	16.3
50	150	21.9	17.9
45	166	24.3	19.9
40	188	27.4	22.4
35	214	31.2	25.7

Source: Compiled by the authors.

TABLE 5.2

1973-74 Capacity Factors by Age of Plant for Boiling Water and Pressurized Water Nuclear Reactors

Years of Service	No. of Plants[a]	Average Plant Capacity Factor[b] (percent)
1-2	8	54.3
2-3	5	50.6
3-4	5	63.5
4-5	3	61.2
5-7	4	66.7
11-14	3	39.3

[a]Weighted by MWe capacity.

[b]Average capacity factor: 55.2 percent.

Source: David Dinsmore Comey, *Nuclear Power Plant Reliability: The 1973-74 Record*, BP 1-7507, (Business and Professional People for the Public Interest), February 14, 1975, citing data from Nuclear Regulatory Commission Reports on availability and capacities of nuclear power plants.

TABLE 5.3

1961-73 Coal-Fired Plant Capacity Factors vs. Age of Plant

Years of Service	No. of Plant-Years[a]	Average Capacity Factor[b] (percent)
Up to 2	67	59.2
3	58	61.8
4	47	64.4
5	37	61.8
6	32	62.7
7	22	65.3
8	16	65.6
9	11	62.8
10	9	67.6
11	6	66.3
12	5	56.4
14	6	55.5

[a]Plants were excluded from age analysis when they expanded capacity, and the part of the plant that represents the expanded capacity would be of a different age than the rest of the plant. For full methodology see Appendix E.

[b]Weighted by MWe capacity.

Source: Federal Power Commission reports, *Steam-Electric Plant Construction Cost and Annual Expenses*, Annual Supplements Number 14-25 (1961-72). 1973 data on capacity factors was drawn from *Steam-Electric Plant Factors, 1974 Edition* (National Coal Association, Washington, D.C.) December 1974.

our history tells us this is an unrealistic goal. We have kept records on capacity factors on all new base-load units installed in our system since 1938 and we have found that the average capacity factor for the first five years of operation life of the 36 fossil units we installed during that period of time has been about 68 percent. After the sixth or seventh year, capacity factors on all units began to decline as they were replaced by newer, more economical units."[2]

From various sources, the following data can be summarized:

	Capacity Factor (percent)
28 nuclear reactors above 100 MWe[3]	55.2
894 fossil-fired plants[4]	68.9
158 fossil plants above 390 MWe[5]	63.2
68 coal-fired plants above 100 MWe[6]	61.2

A primary method of evaluating the reliability of plants is to categorize them by age of plant as presented in Tables 5.2 and 5.3 for plants larger than 100 MWe.

INTERPRETING THE DATA

Electric utilities in general operate plants to meet three levels of demand: base-load, intermediate-load, and peak-load. The prime function of a base-load plant is to run at full capacity all year long to meet the minimum level of demand that exists around the clock. Intermediate-load plants, which are frequently on line, supplement the output of base-load plants by meeting increases in demand that occur with some regularity; peak-load plants operate only occasionally when demand reaches very high levels, such as on hot summer days when air conditioning units are going full bore.

For this reason, analysis of the significance of capacity factors of fossil fuel plants is particularly difficult because the statistics do not state whether a plant operated at a certain level by design or because of equipment failures. It is common among fossil plants that as newer, more efficient units come on line, the older facilities are converted from base-load to intermediate-load use.

With nuclear plants, however, this interpretation of data does not pose a problem since the design characteristics are such that they do not track load changes well and, therefore, only lend themselves to base-load use. But the one difficulty in analyzing the significance of nuclear data is that the number of plants in the sample size of atomic facilities is statistically small. However, if we limit the comparison between coal and nuclear only to large plants, the inference could be drawn that apples are being compared with apples.

TABLE 5.4

Capacity Factors by Size of Plant

Design Power Level	Nuclear		Coal	
	No. of Plant-Years	Avg. Capacity Factor*	No. of Plant-Years	Average D.F.*
Under 500 MWe	14	61.9	223	68.1
500-1,000	39	54.7	107	62.3
Over 1,000	1	37.8	104	57.8
	Total: 54	Avg. 55.2	Total: 434	Avg.: 61.2

*Weighted by MWe capacity.
Source: Comey and FPC reports.

An advantage of comparing only large plants is that within the past few years the utility industry took a "giant leap forward" and began constructing both coal and nuclear 800 MWe and 900 MWe plants without scaling up gradually from the 300 MWe level. This means that in a sense, new technologies are being compared for both types of plants. This would include the complexities of adding to both categories of plants new equipment such as cooling towers and cooling machinery for generators, as well as coping with the practical running and maintenance problems of very large generators. From the data in Table 5.4, it is evident that size for size, or age for age, coal plants outperform nuclear plants. This can be seen visually in Figures 5.1 and 5.2, which compare both types of energy source.

The proponents of nuclear power claim that in time, the industry's record will improve. Yet, commercial plants have been operating in excess of 15 years, and the relevant questions are: "When do the promises become reality? " and "Should a billion-dollar decision be based on the hope of future improvement? " Some problems unique to the atomic industry would seem to militate against substantial improvements in nuclear power capacity factors to the point where they would surpass those of coal.

FIGURE 5.1

Coal and Nuclear Capacity Factors by Age of Plant

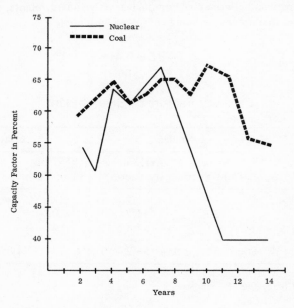

FIGURE 5.2

Coal and Nuclear Capacity Factors by Size of Plant

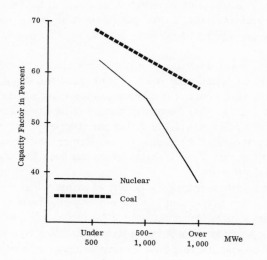

The first of these obstacles is the difficulty of repair under radioactive conditions. To prevent excessive radiation exposure, workers must be suited up and rotated on cycles as short as 60 seconds, which can then become the maximum allowable dose for a three-month period. A typical example was the repair of a cooling pipe at Consolidated Edison's Indian Point plant, which took seven months, 700 men, and cost $1 million. According to Louis Roddis, vice president, a comparable repair in a fossil station would have required only 25 men and would have been accomplished within two weeks.[7]

The second special circumstance that reduces the capacity factors of nuclear plants is their control by the Nuclear Regulatory Commission (NRC), which can order immediate shutdowns of all plants if they feel there is a serious question of safety. This occurred in February 1975, when cracks in cooling pipes of a few reactors were found and all 23 plants of a similar design were forced to close for inspection. Another example is the Fort Calhoun station in Nebraska, where vibrating fuel rods resulted in the plant operating at only 50 percent capacity for a number of months.

The NRC can also order plants to shut down for retrofit modifications in order to meet new and higher safety standards. Whether these actions may or may not be justified is often the basis for heated debate. The fact is that the commission does have the power to take these steps, and does have an obligation to protect the citizenry. In the tug of war between the environmentalists on one side, and the manufacturers and utilities on the other, guessing at the final position of the "happy" middle ground between overregulation and insufficient control can provide endless hours of speculation.

One possible scenario is that as more nuclear plants are built and the economy becomes more and more dependent upon nuclear power, industry pressure will tend to attenuate the regulatory activities of the NRC. Then, after a serious nuclear accident occurs, the pendulum will swing to the opposite extreme, and the capabilities of nuclear plants to produce power will be sharply curtailed.

According to some observers (not necessarily critics) of the nuclear scene, the low reliability of commercial reactors is uncalled for. They point out that the nuclear power plants built for military purposes, such as submarines, have been much more reliable. This lower commercial reliability is not attributed to the greater size of civilian plants, but rather the stricter quality control requirements of military programs. According to one former AEC commissioner, lack of enforcement of quality control requirements has been the "Achilles heel" of the civilian nuclear program.[8]

The problem with increasing the quality of plants is that it could boost the capital cost to such an extent that nuclear power would no longer be cost competitive with other sources. As an example of the ultimate in quality control, the Navy's submarine program under Admiral Hyman Rickover cost $2,400/kilowatt in the 1960s for nuclear reactors built to very stringent quality requirements.[9] (This compares with an approximate $120/kilowatt cost for commercial plants in 1965.) The alternative is to build, as is now done, to that specification which theoretically should perform adequately, and hope that the plant functions according to plans.

While quality control, regulation by the NRC, and problems with repairs in a radioactive environment may be the bugaboos of the nuclear industry, compliance with sulfur emission controls is the bane of coal plants. In order to comply with standards of the Clean Air Act, utilities can take one of five options:

1. Utilize low-sulfur coal.
2. Employ intermittent control coupled with fuel substitution during weather inversions.
3. Disperse pollutants through the use of tall stacks.
4. Mechanically or chemically clean the coal prior to burning.
5. Clean it after combustion with stack gas scrubbers.

The use of low-sulfur coal is preferable if coal is available at an economical price. Intermittent controls and tall stacks are basically unacceptable to the Environmental Protection Agency, and are being employed solely as temporary measures. The cleaning of coal has not yet been commercially demonstrated, which leaves stack gas scrubbers as the only present short-term alternative to burning low-sulfur coal. There is currently substantial disagreement over the use of scrubbers, but EPA maintains that there is sufficient experience to justify use by the utility industry.[10] To support this claim, EPA cites a NATO report,[11] which indicates that scrubber systems have been effective in Japan and Europe, and a recent hearing panel report.[12]

On the other side, the Bureau of Mines, the Federal Power Commission, and the utility industry maintain that sulfur emission technology has not yet reached the stage of commercial development where it is an effective solution to the problem of burning high-sulfur coal. But EPA now reports that newer scrubber systems designed to avoid the earlier mechanical and corrosion problems are operating at better than 90 percent availability.[13] To determine the overall effect upon capacity factors, more experience will have to be gained before a final evaluation can be made. It would appear that a pre-washing process which was not an integral part of the combustion phase would ultimately prove to be a preferable system if it is cost-effective.

Finally, in evaluating the effects of selected variables upon the reliability of plants, historical evidence seems to show that larger plants are less reliable than smaller ones. Whether this was caused by the too rapid scaling up that was discussed previously, or whether there are inherent design limitations due to the properties of materials and the laws of physics, is not evident. However, if large units continue to operate at lower capacities than smaller units, the economies of large scale plants may prove to be illusory.

SUMMARY

At every age level, coal-fired plants have had a better record of performance than nuclear plants, with an overall average of 8 percent higher. Plants over 500 MWe of both types have had lower capacity factors than smaller plants. For nuclear plants, there is an average 10.5 percent difference, and for coal plants, an 8.6 percent spread.

CONCLUSION

Utility executives must begin to integrate realistic capacity factors into their economic decision-making methodology. Based on evidence from published analyses by industry and government, together with applications filed before various state regulatory agencies, the use of overly optimistic capacity factors seems to be a prevalent practice. Just as executives of major trucking companies base the purchase of a fleet not only on initial first cost, but also upon the reliability of a specific type of truck and its ability to be available for service at its design capacity, so must power plants be selected on the basis of overall real-world performance standards.

NOTES

1. Investor Responsibility Research Center, Inc. (IRRC), *The Nuclear Power Alternative*, Special Report 1975-A (January 1975): 17.

2. Ibid., p. 22.

3. David Dinsmore Comey, *Nuclear Power Plant Reliability: The 1973-74 Record*, BP 1-7507 (February 14, 1975).

4. Edison Electric Institute, *Report on Equipment Availability for the Ten-Year Period 1964-1973*, EEI Publication No. 74-57 (New York: December 1974). In this data for the years 1964-73, there is no differentiation between oil, coal, or gas-fired plants, nor between base-load units and others.

5. Ibid.

6. Federal Power Commission Reports, *Steam-Electric Plant Construction Cost and Annual Expenses*, Annual Supplements Nos. 14-25 (1961-72).

7. Des Moines *Register*, October 7, 1974, p. 8.

8. IRRC, op. cit., p. 22.

9. *Congressional Record*, May 15, 1973, Testimony of Captain William Heronemus (former supervisor of nuclear submarine construction for Admiral Rickover).

10. *Final Report, Sulfur Oxide Control Technology Assessment Panel on Projected Utilization of Stack Gas Cleaning Systems by Steam-electric Plants*, April 15, 1973, submitted to the Federal Interagency Committee for Evaluation of State Air Implementation Plans, from FEA, p. 68.

11. *Control Techniques for Sulfur Oxide Air Pollutants*, October 1973, prepared by the Expert Panel for Air Pollution Control Technology, The Committee on the Challenges of Modern Society, NATO: John Burchard, Panel Chairman, from FEA.

12. *Report of the Hearing Panel, National Public Hearings on Power Plant Compliance with Sulfur Oxide Air Pollution Regulations*, January 1974, from FEA, 69.

13. EPA, *Report to Congress on Control of Sulfur Oxides*, EPA-45011-75-001 (February 1975): 31.

6

ANALYSIS OF
BASE YEAR COSTS

A forecast of future costs is only as valid as the accuracy of base point data and the assumptions being used. Therefore, step one is to ascertain as accurately as possible current (or 1975) costs; and step two is to modify these costs so that they will reflect future parameters rather than present conditions. As an example, past and present costs of uranium enrichment to commercial users have been significantly below "true economic cost." The government has subsidized this activity and has utilized a plant that has been substantially depreciated. A private corporation, however, would have had to recover all of its capital costs, which means that were enrichment to be priced on a "true economic basis," at today's prices, the cost would be more than triple what the government is charging. Future prices would then have inflation added onto this more accurate base.

URANIUM COST

Patterson indicates that uranium prices have been rising dramatically; as of March 1975 they were two to three times what they were in early 1974.[1] *Forbes* reported that "one U.S. producer recently offered uranium for delivery in the mid 1980s at a price of $24 per pound plus an annual escalator of about 7 percent, starting [1975]."[2] The magazine also noted that utilities were offering to buy fuel for future delivery in the early 1980s at prices of $27 per pound, and no offers would be accepted.* A base year cost of $24.70 therefore appears to

---

*On July 7, 1976, the *Wall Street Journal* reported that the price of uranium had further leaped to $40 per pound. However, since all other costs used in the book are from

be conservative. This is equal to 1.49 mills/kwh.*

CONVERSION TO UF$_6$

The AEC gave a figure of $1.50/lb. for conversion in 1974, which was the current market price being charged by private companies. Adding 10 percent for inflation should provide us with a usable 1975 cost of $1.65/lb. Utilizing the same formulas for converting uranium prices into mills/kwh, the cost of conversion is 0.10 mills/kwh.

ENRICHMENT

This operation is now being performed only by government-owned facilities at prices of $42.10 to $47.80 per separative work unit. In WASH 1174, the price shown, however, is $75/SWU, since this was considered to be more in line with what would be charged by a private processor.[3] However, this is significantly below what the costs will actually be.

A major cost of enrichment is the capital facility required. *Barrons* reports that a commercial gaseous diffusion plant would cost $3.5 billion in 1976 dollars or $5.6 billion when completed in 1983, assuming 7 percent inflation.[4] This plant would have a capacity of 9 million SWUs and a 25-year life. But this does not take into account the fact that historically the final costs of building nuclear facilities have been off by more than 100 percent. Nuclear reactor costs, as an example, have escalated at an 18 percent compounded rate from 1965, and at a 26 percent rate from 1970 to 1974.[5] These rates are considerably higher than the general inflation rates for the time periods involved.

1975, consistency will be maintained and the $24.70 figure used.

    *Calculation of mining and milling costs per kilowatt-hour:
At 0.30 percent tails assay,

    7.08 kg U$_3$O$_8$ = 1 kg enriched uranium

    1 kg = 2.205 lb.

    7.08 × 2.205 = 15.6114 lb.

    15.6114 lb. × $24.70/lb. = $385.60 (the cost of U$_3O_8$ needed for 1 kg enriched uranium)

    1 kg U produces 258,200 kwh = $0.0014934/kwh or 1.49 mills.

The fact that virtually all estimates of future plant costs are off by a large magnitude is supported by ORO-684.[6] Projections made in 1971 gave costs of $1.2 billion and $1.05 billion (using advanced technology) for a gaseous diffusion plant of 8.75 million SWU. Yet, 1975 estimates were $3.27 billion for 9 million SwU or $3.18 billion for 8.75 million SWU. This is a cost escalation of 2.65 times to 3.03 times in five years. This would hardly lead to a high confidence level in the prices given in capital forecasts.

If we were to assume a modest cost escalation of 20 percent per year, encompassing a general economic inflation rate of 8 percent, we would have to attribute a 12 percent per year cost rise to factors outside of the overall national inflation rate. Over an eight-year period, this 12 percent escalation would increase the plant cost by a factor of 2.21 in 1975 dollars, or to $7.23 billion. Assuming a capital recovery rate of 15 percent, the capital recovery factor* for 25 years is 0.1547 and the calculation is:

$$(7.23 \times 10^9)\,(0.1547)\,/\,(9 \times 10^6 \text{ SWU}) = \$124.28 \text{ per SWU}$$

Electric power is another major cost of enrichment. A total of 2,327 kwh/SWU are required. At 24.0 mills/kwh, the cost of electricity per SWU is $55.85.† ORO-684 projects "other production costs" at $2.20, $4.00, and

---

*See Chapter 4 for an explanation of "capital recovery factors" and Table 7.2 for selected CRFs.

†In ORO-684 (p. 28) Table 7, the AEC gives a range of figures of "specific power kw/SWU/yr." of 0.234, 0.270, 0.278. From WASH 1174-74 (p. 47) a figure of 0.267 can be derived as follows:

$$\frac{\text{Power level}}{\text{Annual SWU output}} = \frac{7.38 \times 10^6 \text{ kw}}{27.6 \times 10^6 \text{ SWU}} = 0.267 \text{ kw/SWU/yr.}$$

There are 8,760 hours in a year and the plant operating capacity is 99.5 percent. Therefore:

$0.267 \times 8,760 \times 99.5\% = 2,327$ kwh/SWU

2,327 kwh $\times$ $0.024/kwh = $55.85 as the electrical power cost per SWU.

In ORO-684 (p. 19) the AEC projects power costs of $16.90 per SWU. Dividing this by 2,327 gives $0.00726 or 7.26 mills as the cost of each kwh of electricity. Escalating this by 50 percent for inflation gives 10.89 mills, which is below that of either coal (15.6 mills) or nuclear (12.2 mills) as given by WASH 1174 (Table 1.5 discounted to 1975). As will be seen, both these costs are inaccurate.

$5.40 per SWU depending upon the power output. For fiscal year 1971, actual "other production costs" per SWU at Oak Ridge, Paducah, and Portsmouth were $8.95, $3.11, and $11.84 respectively. Since one does not know what the economies of scale may actually be, the figure of $5.92 per SWU—the average of all six figures—plus a 30 percent escalation, gives $7.69 per SWU.

The total expected costs for production on a commercial "economic" basis were it to have been performed in 1975 would have been:

Capital costs	$124.28
Electric power	55.85
Other production costs	7.69
	$187.82 per SWU

This is equivalent to 2.66 mills per kwh.*

RECONVERSION AND FABRICATION

This stage is now being performed by private industry and the $70/kg U in 1974 was close to actual market. This will merely be increased to $77 for a 1975 base price, or 0.30 mills/kwh.†

SPENT FUEL SHIPPING

The cost of $10/kg U has been an average cost of shipping to the GE plant at Morris, Illinois, but this plant will not be available in the future. *Barrons* reports that "Tri-State Motor Transit Company, which has specialized most in hauling nuclear materials, finds that one shipment which used to cost about $4,000 for the round trip, now runs to $9,000"[7] because of new security

---

*At 0.30 percent tails, 3.65 kg SWU produces 1 kg enriched U.

3.65 × $187.82 = $685.54/kg U

$$\frac{\$685.54/\text{kg U}}{258,200 \text{ kwh/kg}} = \$0.002655 \text{ or } 2.66 \text{ mills}$$

†($77)/(258,200 kwh) = $0.000298 or 0.30 mills.

regulations. This is an increase of 2.25 times. Using $10/kg as the base and adding 10 percent for inflation yields a cost of $24.75/kg U, or 0.10 mills/kwh.*

REPROCESSING

The $100/kg figure of the AEC in WASH 1174-74 (p. 20) is merely a "best estimate" since reprocessing has never been done on a commercial scale. A rough calculation, however, does not appear to support this number at the present time. The *Atlanta Journal and Constitution Magazine* of February 2, 1975, stated that the Barnwell reprocessing plant would cost an estimated $500 million when completed and would employ 600 workers at an average rate of $11,000 per year. The design capacity is 1,500 metric tons annually. Assuming a capital recovery rate of 15 percent for 25 years, the calculations are:

Capital costs = ($500 $\times$ 10^6)(0.15976) / (1.5 $\times$ 10^6 kg) = $53.25/kg

Labor costs = (600 $\times$ $11,000) / (1.5 $\times$ 10^6 kg) = $4.40/kg

Electric cost = $68.32/kg†

Total cost of reprocessing = $125.97/kg

This is equivalent to 0.47 mills/kwh.‡

---

*($24.75) / (258,200 kwh/kg) = $0.00009585 or 0.10 mills

†A power plant of 750 MWe is required for a 1,500 MT reprocessing plant; assume a 65 percent capacity factor.

750 MWe = 750 $\times$ 10^3 kwh

(750 $\times$ 10^3 kwh) (8.76 $\times$ 10^3 hours/yr.) (0.65) = 4,270 $\times$ 10^6 kwh/yr.

(4,270 $\times$ 10^6 kwh/yr.) / (1.5 $\times$ 10^6 kg/yr.) = 2,847 kwh/kg

2,847 kwh/kg $\times$ $0.024 kwh = $68.32/kg

‡For a year's operation of a 1,000 MWe plant at a 65 percent capacity factor, 22,052 kg of enriched uranium will be required. At the end of the year, this will have decreased in weight by 3.5 percent to 21,280 kg.

21,280 kg U $\times$ $125.97/kg = $2,680,641 as the total cost of reprocessing.

The number of kwh generated is:

WASTE MANAGEMENT

Barrons indicates that the plutonium in spent fuel is "the industry's most troubled segment. . . . it costs $500 per year per kilogram in storage costs (before the new AEC regulations) plus about $1,300 annually in interest. . . . a 1,000 megawatt nuclear reactor will produce 250 kilograms of plutonium per year."[8] Since new AEC regulations increased transportation costs by 2.25 times, we have applied the same factor to plutonium storage.

$500 × 2.25 = $1,125

(250) ($1,125 + $1,300) = $606,250 storage costs per 1,000 MWe plant, or 0.11 mills/kwh.*

Nuclear Fuel Cycle Costs

Cost Component		Cost: mills/kwh
Mining and Milling ($24.70/lb. U_3O_8)		1.49
Conversion to UF_6 ($1.65/lb. U)		0.10
Enrichment ($187.82/kg SWU)		2.66
Reconversion and fabrication ($77/kg U)		0.30
Spent fuel shipping ($24.75/kg)		0.10
Waste management ($606,250/1,000 MWe plant)		0.11
Fuel inventory carrying charge (at 15 percent)†		0.71
Total cost		5.47 mills/kwh
Reprocessing‡ ($125.97/kg U)	0.47	
Mining and milling not required	(0.29)	
Conversion to UF_6 not required	(0.02)	
15 percent carrying charge	0.02	
Total reprocessing cost	0.18	
Total fuel cycle costs		5.65 mills/kwh

(8.76 × 10³ hrs./yr.) (10⁶ kwh plant size) (0.65) = 5.69 × 10⁹ kwh/yr.

Therefore,

($26.80641 × 10⁵) / (5.694 × 10⁹ kwh) = $0.000470783 or 0.47 mills/kwh
*($6.0625 × 10⁵) / (5.694 × 10⁹ kwh) = $0.0001064 or 0.11 mills.
†This is an AEC figure.
‡Reprocessing may not be economical for a utility, but the limited supply of uranium ore makes this step a necessity. The theory is that reprocessing should account for

COAL FUEL COSTS

From Chapter 3, fuel costs for Des Moines, Iowa would be:
5.43 mills for Wyoming strip-mined coal.
8.45 mills for Illinois deep-mined coal.

OPERATING AND MAINTENANCE COSTS

In WASH 1174, O & M costs of 0.81 mills for nuclear and 1.999 mills for coal are given.[9] The Investors Responsibility Research Center (IRRC)[10] reports that a survey of Federal Power Commission data on 20 large nuclear plants operating in 1973 showed that five had O & M costs greater than 2.0 mills/kwh, nine between 1.0 and 2.0 mills, and six less than 1.0 mill/kwh. The coal figure of 2.0 mills appears to be valid for newer coal plants with flue gas scrubbers used for the removal of excess sulfur. If the IRRC data is averaged, the cost calculated for nuclear plants is 1.475 mills.

CONSTRUCTION COSTS OF NUCLEAR AND COAL PLANTS

Trying to determine the construction costs for a large power plant in any one year is more difficult than it would appear at first glance. A television set, for example, may be built within the short span of a few days from the time parts are received to the day the completed set comes off the assembly line. But a base-load power plant is built over a six- to 10-year period. Thus, should the "current cost" selected for any given year be the cost of a plant *completed* or *begun* in that year?

In *Power Engineering*, senior editor F. C. Olds observed[11] that from 1965 to March 1, 1975, power plant costs had risen at a rate of 18 percent per year compounded, and from 1970 at an annual rate of 26 percent. Olds then developed a "moving target" type of methodology for getting around this time span problem, and derived a smoothed trend line that omitted some aberrations and resulted in a figure of approximately $500 per kw as the cost of a nuclear

---

something on the order of 20 percent of the feed material that is required by a utility. Thus, on a 0.30 tails assay basis, the utility will receive a credit of 1.4 kg on each 7.1 kg of feed of U required. (At the enrichment stage, reprocessed feed is mixed with natural feed.) Thus, the credit is presented as part of the reprocessing stage since without reprocessing, the fuel cycle cost above (which is based on 7.1 kg being natural feed) applies. In actual practice, there is no reprocessing being performed currently and the percentage of fuel that is reprocessed will probably be much less than 10 percent. This calculation, however, is performed in line with ERDA's current doctrinaire computational methodology.

plant during the first quarter of 1974. Olds continues to monitor the escalation of costs and when asked for his estimate of what the cost of a nuclear plant would be if one could be built "overnight" in mid-1975, he said, "If I were going to do it, I'd take $650" as the per kilowatt figure.[12] Another approach—given in Appendix D—derives a figure of $633, which closely fits the Olds estimate.

Regarding coal plants, Olds wrote, "Some soundings of the technical community produced some evidence that a modern coal-fired plant today would cost about 75 percent as much as its nuclear counterpart."[13] Other computations in the article show that modern coal plants with scrubbers and cooling towers might cost 82 percent of what a nuclear plant costs. If $640 is used as the mid-1975 "overnight" cost of a nuclear plant and $525 (82 percent of $640) as the cost of a coal-fired plant, with a 30-year life, a 65 percent capacity level, and a capital recovery rate of 15 percent, the following calculations can be performed.*

8,760 hrs./yr. $\times$ 0.65 = 5,690 hours of use

Nuclear: ($640 $\times$ 0.15230†) / 5,694 = $0.01712 or 17.12 mills/kwh

Coal: ($525 $\times$ 0.15230) / 5,694 = $0.01404 or 14.04 mills/kwh

Cost of Electricity for Des Moines, Iowa (Mills/kwh)

	Nuclear	Western Coal (Strip-Mined)	Illinois Coal (Deep-Mined)
Capital	17.12	14.04	14.04
Fuel	5.65	5.43	8.45
O & M	1.48	2.00	2.00
	24.25	21.47	24.49

THE EFFECTS OF CAPACITY FACTORS

The previous computations assumed that all plants operated at the same 65 percent capacity level. However, in the real world, there have been and will

---

*The figures of $640 and $525 represent an allocation of approximately 75 percent to "brick and mortar" and 25 percent to "cost of money."

†CRF = 0.15230.

continue to be deviations from the target. If the historical averages for nuclear and coal given in Chapter 5 are used, the capacity factors would be adjusted as follows:

Actual Capacity Factors: Nuclear = 54.6 percent

Coal = 62.4 percent

Taking these figures into account, the cost of electricity for Des Moines (in mills/kwh) would be revised from the above table along these lines:

	Nuclear	Western Coal	Illinois Coal
Capital	20.38	14.63	14.63
Fuel	5.65	5.43	8.45
O & M	1.48	2.00	2.00
	27.51	22.06	25.08

The reader, in reviewing other material, will often find certain assumptions regarding capacity factors. Typically, unrealistically high ones will be used which, of course, understate the allocation of plant cost per kwh. The generalized method for converting a given capacity factor to another base is:

$$\frac{\text{given capacity factor}}{\text{new capacity factor base}} = \text{conversion number}$$

Then take the conversion number and multiply it by the capital cost component. Using numbers above as an example:

(65 percent) / (54.6 percent) = 1.1904761 = conversion number

$1.1904761 \times 17.12 \text{ mills}_{nuclear} = 20.38$ mills

Another approach to accounting for capacity factors is to apply probability theory to the computation. Utilizing the data from Table 5.1, a subjective probability of occurrence (normally based on actual historical distributions of frequency of occurrence) is then applied to each capacity level as shown below.

TABLE 6.1

Hypothetical Illustration of the Use of Probability Analysis in Computing Capital Costs

Capacity Factor (Percent)	Capital Cost (Mills/kwh)	Probability	Col. 2 X Col. 3
75	14.6	.05	0.73
70	15.6	13	2.03
65	16.8	.25	4.20
60	18.2	.18	3.28
55	19.9	.12	2.39
50	21.9	.10	2.19
45	24.3	.09	2.19
40	27.4	.06	1.64
35	31.2	.02	0.62
		1.00	19.27 mills/kwh = expected cost

Source: Data compiled by the authors.

There is an additional consideration. When a plant is off-line for an unscheduled period, the utility either must turn on old inefficient plants, or, at a negotiated price, purchase power off the grid—or do both. In either case, the amortization for the plant that is down must still continue. Assuming a cost of 20 mills for the purchased power,* the computation would be:

	Nuclear	Coal
Capital Amortization	20.38	14.63
Purchased Electricity	20.00	20.00
	40.38 mills	34.63 mills

Because of the continuing higher amortization of a reactor, actual cost of downtime will always be higher for nuclear power.

---

*This assumes that the purchaser must pay the marginal cost of the seller which would be the cost of the electricity generated by the newest plant.

SUMMARY

Three major elements enter into the computation of electrical generating costs. For Des Moines, Iowa in 1975, on a true economic basis, the cost of electricity in mills/kwh for a new plan would have been:

	Nuclear	Western Coal (Strip Mined)	Illinois Coal (Deep Mined)
Capital	17.12	14.04	14.04
Fuel	5.65	5.43	8.45
Operating & maintenance	1.48	2.00	2.00
Total (mills/kwh)	24.25	21.47	24.49

When analyzing the comparative costs of electrical generation, bear in mind that generating costs comprise only 25 to 40 percent of the amount charged a consumer. Therefore, a 20 percent difference in generating cost may only amount to a 7 percent spread in the final retail price.

CONCLUSION

Especially for coal facilities, the difference in fuel costs will vary considerably from one geographic region to another because of the impact of freight costs. Creative management, however, can often minimize these costs as Detroit Edison plans to do with Montana coal, which is being partially shipped by boat. Another alternative is to improve the technology of electrical transmission so that line losses over long distances will be negligible. If this is accomplished, coal burning plants built adjacent to the mines reduce the transportation cost to zero. Another option is to build a coal plant directly adjacent to or within a city, then install a steam distribution system and sell the steam that is generated as heat for buildings. This approach to better plant utilization and energy conservation is being actively pursued in Germany.

In the final analysis, a decision cannot be made simply by calculating to the third decimal place the cost of producing power and then choosing the lowest figure. Generating costs typically account for only 25 to 40 percent of the price a consumer pays. In 1972, the average residential revenue per kwh for Iowa Power and Light Company was 26.2 mills, but the cost of producing power from the company's Council Bluffs plant was less than 10.0 mills.

While coal may be fractionally cheaper for an Iowa utility, additional freight costs (because of greater distances from coal mines) may cause coal-fired plants to be more expensive for an eastern utility. Nonetheless, in making a decision on a power source, the utility must also consider the element of risk. If the superficially cheapest power is not the most reliable, then it is clearly no longer the least expensive. And reliability analysis must encompass not only the functioning of a plant (for which an accurate capacity factor forecast is a necessity), but the entire system's risk factors as well. A significant risk difference between nuclear and coal is the degree of control exercisable by a utility over the fuel cycle. For nuclear, a utility is subject to the vicissitudes of mining companies and cartels, enrichment plants, fabricators, and especially the capriciousness of the federal government. In contrast, for coal, the air pollution ground rules have already been established, and the utility can exercise virtual control over every aspect of the coal cycle except the condition of railroad tracks. Only when risks are accurately analyzed and included in the company's analysis can a true "economic" decision be reached.

NOTES

1. John A. Patterson, Chief, Supply Evaluation Branch, Division of Production and Materials Management, ERDA, *U.S. Uranium Situation*. Speech given at the Atomic Industrial Forum Fuel Cycle Conference 75, March 20, 2975, p. 2.

2. *Forbes*, "It worked for the Arabs . . . ," January 15, 1975, p. 20.

3. WASH 1174-74, op. cit., p. 20.

4. Charles Joslin, "Uranium Enrichment," *Barrons*, July 7, 1975, p. 11.

5. F. C. Olds, "Power Plant Capital Costs Going Out of Sight," *Power Engineering*, August 1974), p. 36.

6. U.S. Atomic Energy Commission, *AEC Gaseous Diffusion Plant Operations*, ORO-684 (January 1972): 28.

7. Charles Joslin, "Nuclear Genie," *Barrons*, September 23, 1974, p. 11.

8. Ibid.

9. Table 1-5 (p. 6) discounted to 1975.

10. IRRC, op. cit., p. 24.

11. Olds, op. cit., p. 37.

12. Olds, Telephone conversation, July 30, 1975.

13. Olds, op. cit., p. 43.

**PROJECTING
FUTURE COSTS**

The previous chapter developed the "true economic cost" of both nuclear and coal power as of 1975, which can serve as the starting point for projecting the future. This chapter will present a "cookbook" model for developing future costs, with the ability to use different capital recovery rates and also different rates of inflation for each individual element. The chapter consists almost entirely of mathematical equations and may be skipped by those not wishing to actually calculate any projections. The example that will be used is a nuclear plant beginning construction in 1980, with completion at the end of 1989, and going on line in 1990.

The capital cost equation is:

Amortized annual capital costs in mills/kwh = Y

$$= \frac{(B)\,(1 + I_1)^{n_1}\,(C_1)\,(R)\,(1{,}000)}{(8760)\,(A)}$$

Where:

B = Base year plant cost in \$/kw

I_1 = Inflation rate for capital costs

n_1 = Number of years from base year to start of construction

C = Construction factor taken from Table 7.1 (or individually derived as shown in Appendix D)

R = Capital recovery factor, a function of the selected capital rate and the expected economic life of the plant, obtained from "Capital Recovery Tables, Uniform Series," excerpts of which are given at the end of this chapter in Table 7.2

1,000= Conversion of dollars into mills

8,760= Hours in a year

A = Capacity factor

Example:

B = $640

I_1 = 15 percent

n_1 = 5 years

C_1 = 2.41902

R = 0.1523 for 15 percent/30 years

A = 65 percent

$$Y = \frac{(640)(1 + 0.15)^5 \ (2.41902)(0.1523)(1,000)}{(8,760) \ (0.65)} = \frac{474,152}{5694}$$

= 83 mills/kwh

For fuel and operating costs, the equation is:

$$F_1(1 + I_2)^{n_2} + M_1(1 + I_3)^{n_2} = U$$

Where:

F_1 = Fuel cost in mills/kwh for base year;

I_2 = Inflation rate for fuel

n_2 = Number of years from base year to year of operation in mills/kwh

M_1 = O & M cost for base year

I_3 = Inflation rate for O & M costs.

Example:

F_1 = 5.93 mills/kwh

I_2 = 20 percent

n_3 = 15 years

M_1 = 1.48 mills/kwh

I_3 = 12 percent

U = $(5.93) (1.2)^{15} + 1.48 (1.12)^{15} = 91.36 + 8.10 = 99.46$ mills/kwh

Total costs:

Capital	83 mills/kwh
Fuel and O & M	99
Total	182 mills/kwh

The same method applies to coal, of course.

TABLE 7.1

Construction Factors

Inflation Rate (percent)	Nuclear Factor	Coal Factor
6	1.44	1.31
8	1.62	1.43
10	1.82	1.56
12	2.01	1.69
15	2.42	1.92
20	3.19	2.35
25	4.16	2.86
30	5.38	3.46

Note: The methodology for deriving these factors is presented in Appendix D.
Source: Compiled by the authors.

TABLE 7.2

Selected Capital Recovery Factors

	Number of Years			
	25	30	35	40
8 percent	0.09368	0.08883	0.08580	0.08386
10 percent	0.11017	0.10608	0.10369	0.10226
12 percent	0.12750	0.12414	0.12232	0.12130
15 percent	0.15470	0.15230	0.15113	0.15056
20 percent	0.20212	0.20085	0.20034	0.20014
25 percent	0.25095	0.25031	0.25010	0.25000

Note: The generalized formula is:

$$[i(1 + i)^n] / [(1 + i)^n - 1]$$

where:

i = interest rate

n = number of years.

Source: Compiled by the authors.

Due to the scarcity of uranium resources, one hope of the future is the development of the liquid metal fast breeder reactor (LMFBR), which can utilize nonfissionable U^{238}, produce heat to create electricity, and simultaneously convert the uranium to plutonium (Pu^{239}), which can then be utilized to fuel either conventional reactors or additional breeder reactors.

The government is scheduled to begin construction on the Clinch River Breeder Reactor (CRBR), a 350 MWe plant. Commercial operation is scheduled to begin in 1983, but this date has been slipping each year, and will probably not be realized.* With conventional nuclear plants requiring a nine- to 10-year lead time, a seven-year timetable for a new venture would appear to be overly optimistic. The cost of the CRBR has escalated measurably and the ultimate expected cost is difficult to determine. In August 1972, estimates were $700 million.[1] By June 1974, this had climbed to an ERDA estimate of $2.1 billion, of which $1.68 billion is attributed to escalated capital costs and the balance for further development costs, fuel cycle operation and maintenance, and costs for a five-year demonstration period.†

Assuming that future estimates do not increase any further, this is equivalent to $4,571/kw in capital costs, which compares with an expected cost of

---

*By June 15, 1976, the Federal Energy Resources Council in its report "Uranium Reserves, Resources, and Production" wrote: "Current ERDA plans call for demonstration of the breeder as a safe, reliable, and environmentally sound energy producer by 1987."— a four year slippage during the one-year period 1975 to 1976.

†ERDA estimated 1974 base capital costs at $1.12 billion and assumed 8 percent per year escalation.

$1,548/kw for a conventional light water reactor (LWR) completed in 1984.* However, the ERDA estimate for the breeder may be too low, as only an 8 percent per year escalation rate was utilized; yet utility construction costs, as stated earlier, have actually been rising at an 18 to 26 percent annual rate. The final cost of any breeder plant can only be speculated upon. But after 22 years of conventional nuclear plants, costs have continually outpaced the inflation rate by a minimum of two times.

Using a 65 percent capacity factor and a 15 percent capital recovery rate, the $4,571/kw and $1,545/kw figures translate into generating costs of:†

	LMFBR	LWR
Capital	122.2	41.4
Fuel	—	19.9
O & M	5.2	5.2
	127.4 mills/kwh	66.5 mills/kwh

Whether the high breeder cost is solely attributable to the research and development expense of an experimental program, or will be representative of LMFBR costs in production, cannot be determined at this time. But we do know that prior AEC estimates have been continually understated by large multiples. *Science* magazine wrote: "Whether because of the complexity of the technology or the AEC's unrealistically low estimates, cost overruns have been endemic to the breeder program. A major test reactor, the Fast Flux Test Facility (FFTF), now being constructed in Hanford, Washington, rose from $87 million to $450 million, and the program as a whole has jumped from $2 billion to more like $5 billion. . . ."[2]

A selling point of the breeder is that it generates more fuel than it uses, but the annual quantity produced is not very great. It will take a breeder reactor 15 to 20 years to breed enough plutonium for the core of a second breeder, and longer still to breed the "pipeline inventory" required for the fuel cycle of the second breeder.[3] At this time, ERDA does not project that the LMFBR will make a substantial contribution to the energy picture even by the year 2000. The current projections are:[4]

---

*Assuming $640/kw base cost in 1975, and 15 percent per year escalation.
†Assumes 15 percent per year escalation on all costs.

TABLE 8.1

LMFBR Power Capacity in Megawatts

	1985	1990	1995	2000
Low	350	1,150	7,000	61,000
Moderate/Low	350	1,150	8,100	81,100
Moderate/High	350	1,150	9,000	100,000
High	350	1,150	11,400	125,000

Source: U.S. Energy Research and Development Administration, Office of the Administrator for Planning and Analysis, *Total Energy, Electric Energy, and Nuclear Power Projections*, February 1975.

By the year 2000, these estimates would represent only 10 percent of all forecast nuclear capacity, and between 3.9 and 5.7 percent of total electrical capacity, depending on which case is referenced. Currently, the LMFBR faces uphill sledding against environmentalists both within and outside government, with heavy opposition developing over the plutonium produced in the course of operation, and the plans for the subsequent conversion of the plutonium into fuel.* In mid-1974, the AEC submitted a 2,200-page analysis of the environmental effects of the LMFBR, and the Environmental Protection Agency gave the report its lowest rating of "3," indicating that the analysis was "inadequate" and in need of "substantial revision." *Science* wrote:

> Among its major points, the EPA said that the AEC provided vague and mostly qualitative indications of its approach to major problems of reactor safety; that it provided no assurance that plutonium fuel could be protected from theft at an acceptable cost; and that the volume of wastes produced by large numbers of breeders may have been underestimated.
>
> Most of the EPA's criticism, however, centered on the commission's optimistic analysis of the breeder's economic cost and benefits. The EPA points to half a dozen technical flaws or omissions, all of which have the effect of either inflating the projected benefits or minimizing the costs.[5]

---

*This subject was touched on in Chapter 2.

Even the nuclear industry itself has been wavering in its support of the breeder program, as reported by *Science*: "Even from the nuclear industry's professional press some discordant notes on the breeder can be heard. *Nuclear News* (August 1974, p. 55), journal of Nuclear Society, has published an article sharply critical of the breeder program and its present goals."[6]

The LMFBR may or may not become a reality and solve the problems of insufficient uranium resources at some point after the turn of the century. Presently, it will "play little or no role as a short-term energy option in Project Independence"[7] and its future will be resolved only with the passage of time. For now, it should be considered solely as experimental, rather than a viable economic alternative to either coal or conventional nuclear power. Perhaps an even more important consideration is the future role of solar power, which may be cheaper than breeder power. Since solar energy entails no environmental problems, utilizes a significantly less complex technology, and has a very low investment risk, the breeder may be obsolete before it emerges from the research and development stage.

NOTES

1. General Accounting Office, *Report to the Congress: Cost and Schedule Estimates for the Nation's First Liquid Metal Fast Breeder Reactor Demonstration Power Plant*, May 22, 1975.

2. "Complications Indicated for the Breeder," *Science*, August 30, 1974, p. 768.

3. John Price, *Dynamic Energy Analysis and Nuclear Power* (London: Friends of the Earth Ltd. for Earth Resources Research Ltd., December 18, 1974), p. 22.

4. ERDA projections given in Appendix A.

5. "Low Marks for AEC's Breeder Reactor Study," *Science*, May 24, 1974, p. 877.

6. *Science*, August 30, 1974, p. 768.

7. Ibid.

9

CAN UTILITIES SURVIVE MASSIVE CONSTRUCTION PROGRAMS?

As with so many investments and economic forecasts, there is an inclination to project current conditions on a straight-line basis into the future. Seen from the forecaster's point of view, positive trends tend to grow rosier at an ever-increasing rate while negative conditions seem to deteriorate in an accelerated downward slope. The epitome of this thinking is no doubt represented by the stock market. When the market is two years into a bull phase, it appears that there is no top. Conversely, at the end of a bear market, most swamis foresee only the Four Horsemen of the Apocalypse. From a long-term standpoint, this total preoccupation with current conditions tends to obscure a disease that may be eating away at the very entrails of an industry.

SEEING THE FOREST, NOT THE TREES

Theodore Levitt in his book *Innovations in Marketing*, addresses the question of economic metamorphosis rather directly. Although written 13 years ago, his words seem particularly relevant today.

> *Electrical Utilities.* This is another one of those supposedly no-'substitute' products which has been enthroned on a pedestal of invincible growth. When the incandescent lamp came along, kerosene lights were finished. Later the water wheel and the steam engine were cut to ribbons by the flexibility, reliability, simplicity, and just plain easy availability of electric motors. The prosperity of electrical utilities continues to wax extravagant as the home is converted into a museum of electrical gadgetry. How can anybody miss by investing in utilities, with no competition, nothing but growth ahead?

But a second look is not quite so comforting. A score of non-utility companies are well advanced toward developing a powerful chemical fuel cell which could sit in some hidden closet of every home silently ticking off electric power. The electric lines that vulgarize so many neighborhoods will be eliminated—so will the endless demolition of streets and service interruptions during storms. Also on the horizon is solar energy, again pioneered by nonutility companies.

Who says that the utilities have no competition? They may be natural monopolies now, but tomorrow they may die natural deaths. To avoid this, they too will have to develop fuel cells, solar energy, and other power sources. To survive, they themselves will have to plot the obsolescence of what now produces their livelihood. [1]

THE VICIOUS CYCLE

The price elasticity of electricity is not known over a curve with any significant cost rise; that is, we do not know how much consumers will cut back on their usage of electricity if the cost rises past certain price levels. High fuel and construction costs are now conjoining to sharply increase the retail price of electricity. If the dollar value of plant additions doubles over the next five years, but capacity increases by only 25 percent, it becomes quite apparent that the portion of an electric bill represented by capital will rise 60 percent.*

The retail price increases that may be necessary to maintain the solvency of private utilities can lead to what is perhaps the one single circumstance that could universally bring about deterioration in the structure of the private utility industry as it is now configured—permanent excess generating capacity. As demand falls, plant amortization would have to be spread over fewer units of electricity or over less profitable wholesale sales. Rates would rise, and demand would further slacken as alternative sources of energy become cost effective.

While construction and fuel costs of power plants would continue to escalate, the costs of solar and wind power would conceivably decrease as research improved the design efficiency of units and mass production techniques lowered manufacturing costs. In the not too distant future when pollution-free sources of energy have payback periods of less than eight years,† they will then

---

*Some abstract units demonstrate how a doubling of the rate base, but an increase in capacity by 25 percent results in a 60 percent cost increase.

	Cost	Output	Cost/Unit
Old Base	100	100	1
New Base	200	125	1.6

$$†Pay\text{-}back\ period = \frac{(Investment)}{(Savings\ per\ year)}$$

be price competitive and economically attractive to consumers as viable alternatives.*

SHORT-TERM AND INTERMEDIATE-TERM RISKS

While increases in the cost of generating power are inevitable, certain management decisions can have the effect of posturing a company in such a manner as to expose it to an acceleration of the process. There are two kinds of economic risks that can increase the probabilities of encountering financial problems: 1) the building of very large (1,000 MWe) plants (as opposed to smaller units) and 2)the building of nuclear reactors. While large fossil plants may experience some of the problems listed below, they are more apt to occur with nuclear units.

- More exposure to site and vendor strikes.
- Higher probability of larger cost overruns.
- Higher cost of capital as a percentage of total construction costs.
- More exposure to the vagaries of the capital markets with the inherent problems of not selling equity or long-term debt as needed.
- Greater cash flow problems because of exclusion of plant from rate base for a longer period.
- Historically lower reliability of any large plant with concomitant lower capacity factors than smaller plants.
- More severe financial effect from temporary plant downtimes, as higher amortization must still be charged while outside power is purchased.

In addition, there are a number of risks peculiar to nuclear plants:

- Longer lead times than for coal plants, thus accentuating the capital problems cited above.
- Lack of fuel because of inadequate enrichment capacity.
- Plant shutdowns due to insufficient reprocessing capability.
- Highly expensive load curtailment orders caused by Nuclear Regulatory Commission action.
- Greater environmental pressures related to low level radiation emissions and high level waste disposal.
- Potential outcry demanding the closing of all nuclear plants if one serious accident occurs.

---

*See Appendix F for a probable economic scenario of solar power.

While a utility can pass along all the costs incurred with shutdowns once the plant is completed, any such untoward event will raise utility prices for power and speed up the long-term deteriorating downward cycle discussed earlier.

In assessing the possible long-term effects of huge construction programs, it is necessary to differentiate between prudent and imprudent management. Some of the same ills can befall both classes, but the latter category of executive will have created an environment that can hasten adverse conditions, snowballing them into disastrous losses for the company's security holders. The prudent executive would seek to minimize his risks. He would conceivably build coal plants of smaller sizes, which offer the following advantages:

- The lead times are shorter.
- The potential for money market problems is reduced.
- Fuel supplies can be assured on long-term contracts.
- There are no fuel cycle risks.
- Environmental opposition is less vociferous.

THE VIEWPOINT OF A REGULATORY AGENCY

The risks involved in large-scale power plants are now of concern to regulatory agencies, and are being assessed by these bodies, as is evident from the August 19, 1975 Iowa State Commerce Commission "Statement of Intent," which demonstrates their concern with the potential financial problems that can arise from the building of large electric generating facilities. According to the commission:

> . . . the rate level inquiry is in and of itself one of immense concern to this commission and the residents of Iowa.
>
> With the investment magnitude of generation facilities, should a significant part of such investment be excluded from rate base as being an imprudent investment, the impact could very well be destructive to the ability of the utility involved to attract capital or could even result in bankruptcy.
>
> Illustratively, as early as 1968 in *Re Consolidated Edison Co.*, 73 P.U.R. 3d 417, the New York Public Service Commission entertained testimony and argument on the exclusion from Consolidated Edison's rate base of two-thirds of the cost of a nuclear generating facility. While the NYPSC ultimately found the facility to be a totally justified and prudent investment, the case and subsequent cases have underscored the potential impact of an adverse decision. No evidence has been presented to date to support the exclusion from rate base or cost of service as an imprudent expenditure any

investment by any Iowa utility in nuclear facilities. However, the possibility that some such evidence could be proferred at some future date is always present.

If coal curtailment orders and safety incidents continue, the possibility of the investor having his equity investment in utility plant wiped out must inevitably affect his appraisal of the risk attendant an investment in the utility, and hence, increase capital costs to the utility and its customers.[2]

THE POTENTIAL FOR A BAILOUT

Widespread utility cash problems may prompt governments to take a less sanguine view towards rescuing security holders. While Consolidated Edison and Georgia Power were rescued by governmental action, the Penn Central, which was just one more tale of woe in a long history of railroad mismanagement and bankruptcies, has cost investors hundreds of millions of dollars. The errors in judgment that result in plant shutdowns because of nuclear fuel cycle shortages, or because of plants built atop faults (i.e., Con Edison's Indian Point #3 facility) may become the straws that break the back of the proverbial camel; security holders may be sacrificed to the "welfare" of consumers.

MIRRORS AND MICKEY MOUSE SOLUTIONS

Should solar units for individual installation become cost effective, and should construction and capital costs for central stations continue to rise, utilities may more and more begin to look to government schemes for aid. One possibility is the building of power plants by governmental corporations, which sell tax-exempt bonds (and the private utilities would deliver the power). But this is simply a sophisticated version of musical chairs. The customer who buys his power from a completely private utility is, in effect, subsidizing those persons who have public power, since somebody must pay the taxes that are lost (in theory, anyway). The same reasoning holds true for new tax breaks or tax incentives. This is merely tax gymnastics, which moves the tax burden to another sector of the economy in order to support an industry that may no longer be economic in a capitalist sense.

CONCLUSION

There is no substitute for the law of supply and demand. If the cost of one type of energy becomes exorbitant, this will encourage innovative new solutions, which have been the hallmark of U.S. economic strength for most of its history.

Soon after the Civil War, the price of whale oil soared to new highs, just as petroleum has done recently. Then along came cheap kerosene as a substitute for fueling oil lamps. This cycle can repeat itself.

Creative thinking by executives may both minimize risks and find the silver lining in the cloud. To ameliorate financial pressures, utilities could strongly encourage conservation, both through public relations and through lobbying for conservation laws. It may seem odd that a businessman should seek to reduce his revenues, but if he is faced with the options of either excessive risk or the alternative of financial stability, he may opt for the less adventurous course of securing the firm's assets.

Harking back to Theodore Levitt's observations, management officials should ask themselves: "What business am I really in? Am I in the central generating business or the energy business?" One major California utility has evidently decided on the latter—to broaden its scope and, by doing so, prevent its own obsolescence. The utility is promoting the sale of solar units. Most state laws do not prevent utilities from expanding into other fields, and this course of action may not only save utilities from insolvency, but may transform them into "glamor" industries.

NOTES

1. Theodore Levitt, *Innovation in Marketing: New Perspectives for Profit and Growth* (New York: McGraw-Hill, 1962), pp. 43-44.

2. State of Iowa, Iowa State Commerce Commission, *In the Matter of Proposed Construction of Major Utility Plant, Statement of Intent*, Docket No. RES 75-1 (August 19, 1975).

10

"What's the bottom line?" is a question often asked by a corporate board of directors at the end of a presentation in which the speaker has equivocated. The expression is drawn from an analogy with an income statement, where the profit or loss is shown on the bottom line. Now, in a manner of speaking, we have come to "the bottom line" of this study, and the inevitable question must be raised: which energy source is the "best" investment?

We have sought to assemble as fully and completely as possible a fair presentation of the economic pluses and the minuses of nuclear and coal power. It may appear that nuclear power was "slighted" in favor of coal, but, on the contrary, the estimates of nuclear costs were on the conservative side. For example, a 15 percent capital recovery factor was used instead of 18 percent, which would have been more disadvantageous to nuclear because of its higher capital costs. There is also a serious question of whether or not nuclear plants are actually achieving the ERDA conversion figure of 258,200 kwh of electricity for each kilogram of enriched uranium. Some empirical evidence to date indicates that the true ratio may actually be half this amount, which would then effectively reduce the capability of uranium reserves to support generating capacity to half the amount given in Chapter 1.

In determining the costs of coal power, the heat values utilized were lower than the averages given by the Bureau of Mines. This obviously understated the generating capacity of the coal reserves and overstated the cost. In addition, a low recovery rate of 50 percent was utilized in computing the lifetime of coal reserves. In attempting to project the solar costs given in Appendix F, figures that appeared to be the highest feasible were used; for waste energy in Appendix G, 1973 costs were inflated by a generous 50 percent while the savings were understated.

It should also be understood that, much as an effort was made to develop "true economic costs," the criteria for establishing nuclear and coal costs were still not absolutely parallel, since the cost of coal energy, particularly in the capital equipment area, now encompasses "social" costs such as the reduction of pollutants. On the other hand, nuclear power has not accomplished this to date. Specifically, each utility privately pays for only a fraction of its potential accident liability while the balance is covered by the government under the Price-Anderson Act, which is being attacked by some critics as having a limit equal to only 3 percent of the potential damage that could occur. As greater quantities of plutonium are shipped, and tighter transportation and plant security regulations imposed, these costs will tend to escalate. Finally, the true costs of adequate waste disposal are mere guesses at this time. If the government bears the brunt of storing these wastes in perpetuity, then a hidden electricity tax will have been imposed.

A number often bandied about as the total government funds spent on nuclear power is $30 billion. Up to 1976, nuclear plants generated approximately 520 billion kwh of electricity.[1] This would mean that the electricity tax on nuclear was 5.8 cents per kwh, which compares with a typical residential rate of 4.0¢/kwh. Whether these figures are absolutely accurate, and what percentage should be attributed to the civilian nuclear program rather than the weapons program, probably cannot be determined. But we do know that there has been a huge expenditure on nuclear research and development. In fiscal 1974, ERDA's operating budget for non-weapons nuclear R & D was $1.63 billion. For nonnuclear, it was $92.3 million, meaning that almost 18 times more was spent on nuclear. And in previous years even more miniscule amounts were allocated to non-nuclear R & D. In essence, then a "nuclear tax" has been levied indirectly and is not showing up on electric bills.

If we were to treat the pluses and minuses of power alternatives as a financial balance sheet, we would be hard pressed to find very many credits for nuclear energy. Just about all that can be said in their favor is that nuclear plants do work; they do generate electricity. But the negatives are legion:

- a dangerous uranium scarcity
- potential mining and milling shortfall
- international cartelization
- probability of very high future uranium prices. Given a 5X increase in two years to $40/lb. for U_3O_8, $300/lb. is a distinct possibility in the future.
- enrichment capacity shortfall
- reprocessing that has never functioned commercially
- high economic risks attending construction and operation of enrichment and reprocessing facilities
- waste disposal problem never solved

- long-term reliability of plants still a big question. Potential for corrosion and materials fatigue due to radiation exposure over long periods is still not known on any reliable statistical basis.
- presence of radiation, making repairs more difficult than for fossil fuel plants
- lead times for construction too long
- plants essentially uninsurable from a liability standpoint
- security problems on site and particularly of radioactive materials in transit
- many environmental hazards and much opposition from environmentalists
- very high R & D expenditures paid for by tax dollars
- complex, eight-step fuel cycle over which a utility has ownership of only one step—the power plant (Even that one segment of the fuel cycle owned by the utility—its reactor— is not under its control, but under that of the NRC.)
- enrichment and reprocessing, which require government funds and/or guarantees if expansion of new facilities is to take place

In reviewing the economics of nuclear energy, we also raise questions regarding the non-availability of adequate private liability insurance, the atomic and radioactivity exclusions on all casualty insurance, and the fact that even the government limits its liability to a small percentage of the potential damage that could occur.

Historically, the government has issued insurance only when the risks were too high for private carriers to be able to insure on an economic basis. These categories have included burglary insurance in high-crime areas, and property insurance for structures built on flood plains or adjacent to overflowing rivers. If we adhere to the philosophy that a business enterprise must be economically self-sufficient, then the insurance question surely raises some large doubts about the true economics of nuclear reactors.

The next relevant question is: "What's the bottom line for coal?"

Pluses:

- large reserves available; no fuel shortage
- reasonable cost
- independent of foreign sources
- attractiveness of long-term contracts to mine known coal resources as investments for capital markets
- simpler, three-step fuel cycle—mining, transportation, power plant—all of which can be directly owned by utility
- no new technology required for use
- shorter lead times, providing less vulnerability to inflation and greater flexibility for utility planning
- can be used in conjunction with solid waste

Negatives:

- air pollution problems
- additional engineering work required to satisfactorily solve sulfur control problems
- environmental opposition to strip mining
- requires a national commitment to prevent any bottlenecks in the system

In assessing nuclear power, coal power, or in fact any business enterprise, only an evaluation that examines the activity as a total system is valid. Merely having certain elements operational is insufficient because if all components do not function as designed, the system eventually will either become inoperable or its capabilities will be serverely curtailed.

A commercial airline provides an everyday example. Possession of a jet plane is only the very beginning of offering proper service. There must also be a source of spare parts, and of adequate fuel. There must be airport runways capable of accommodating the aircraft, plus ground crews, a reservation system, and of course, trained pilots. Would any airline purchase an airplane if it thought that a full range of spare parts were not readily available, or if it thought its fuel supply might be cut off or severely curtailed, resulting in a high probability that its fleet might be grounded part of the time instead of being airborne and generating revenue at maximum capacity? Yet nuclear power has been operating for over 15 years on just such a basis—a system incomplete and not fully functional, a system with numerous critically weak links. To select a more mundane example, would someone install indoor plumbing in a house, hook up to a water supply, and proceed to use his toilet if he had no sewer or septic tank in which to dispose of the wastes?

Continued construction of (and reliance upon) nuclear reactors, according to any of the ERDA forecasts, will make the 1973 OPEC action seem comparable to the effect of firing a .22 caliber rifle at a tank. Never in history has there been a plan to have the entire economy of a large industrial nation so dependent upon a technology built on so fragile an economic foundation.

Consider that in 1975 dollars, $614 billion for plant and $62 billion for the support facilities of the nuclear fuel cycle—a total of $677 billion*— is projected to be spent by the year 2000. With inflation factored in at 15 percent, this could amount to $5.8 trillion by the end of the century. This compares with

---

* $(959.9 \times 10^6 \text{ kw}) (\$640/\text{kw}) = \$614.3 \times 10^9$

$(960 \text{ plants}) (\$65 \times 10^6 \text{ support costs/plant}) = \$62.4 \times 10^9$

the estimated $160 billion figure for Vietnam, which merely culminated in double-digit inflation.

Consider this possible scenario. The economy grows significantly dependent upon nuclear power to provide its energy needs, when shortages of enriched uranium develop. Power generating facilities shut down. Brownouts develop. Industry cuts back production and workers are laid off. Recession, maybe depression, follows. The effects upon utilities could be disastrous. As generators down, regulatory agencies have to decide whether or not to allow the closed plants to continue in the rate base. If removed from a rate base, bankruptcy for the company would probably occur. If allowed to remain in, electric rates would rise sharply. By comparison, the Equity Funding and Penn Central debacles will look like flea bites.

On a national scale, the deleterious effects upon the economy would make the perturbations and inflation caused by the Vietnam War pale into insignificance. It is acknowledged that these effects would probably not occur until the mid-1980s, but considering the 10-year lead time for nuclear facilities, investment decisions being made now are crucial.

In that "best of all possible worlds" of Candide's friend, the eternally optimistic Dr. Pangloss, the free market forces of supply and depand operating in a classical competitive economic environment would determine selling price; the cheapest, most available source would win the energy sweepstakes. However, the real world does not conform to idealized economic models; such forces as OPEC price fixing, concentrated marketing and production of resources, environmental controls, federal subsidies and tax benefits, governmental price controls, and utility geographic monopoly combined with standard utility regulatory practice, all combine to undermine the functioning of free market forces.

If the government allowed prices to fluctuate freely, and limited its role merely to antitrust action and taxation to prevent "excess profits" (within the framework of balancing environmental with economic needs), the United States eventually would develop the "best" mix of energy sources. Unfortunately, because of its twin powers of taxation and expansion of the money supply, the government is in a position to subsidize "white elephants" that never would survive for one year within the jungle of competition.

In theory, the concept of private enterprise should apply as fully to the energy sector as it does to the garment industry. If private capital assumes a risk without government assistance, and if the investors succeed, they are entitled to a reward. Conversely, if they fail, they lose their investment. In the real world, however, there are cogent "national" considerations for insuring that the power industry does not fail. But this should not be construed to mean that utility executives and the government should be absolved of making sound business decisions.

If government funds are going to be committed to research and development, they should be channeled into that area most vital to the economy and the security of the United States. It would appear that solar energy (including perhaps windpower research) would fit into this category. One argument recently used in the U.S. Senate for voting against an increase in the funding of solar research was that the extra amount could not be fruitfully utilized. This is where the concept of a "national commitment" becomes important. If, for solar energy research, a Manhattan type project (such as developed the atomic bomb) were funded, and if a liberty ship type program to manufacture solar assemblies were subsequently initiated, not only would the economic and physical security of the United States be increased, but the side benefit of reducing unemployment would be realized.

As we near the end of this study, we can't help but raise the question of whether utility executives have been fully cognizant of the economic risks to their companies from continued building of nuclear reactors that may not be able to function because of fuel shortages, and because of the other negative factors associated with atomic power. Are boards of directors prepared to weather a storm of shareholder suits in the future because of imprudent management now?

An executive will always make decisions based on incomplete knowledge and some risks, but when there are such an overwhelming number of problems and potentially extraordinary risks, we wonder whether a prudent businessman would make an investment in nuclear facilities. In most states, there are fiduciary laws that define investment risk in terms of those that would be taken by a "prudent businessman," and the recent Employee Retirement Income Security Act (ERISA) has addressed itself to the question of investment risk in such a manner that significant amounts of capital are no longer available to many financially sound medium sized and smaller businesses. We question if either an interpretation of most state fiduciary laws or ERISA would consider direct or indirect investment in nuclear plants to be qualified for investment by trusts, insurance companies and pension funds if all of the facts regarding the enormousness of economic risks in connection with nuclear plants were fully understood. Those investment managers holding stocks of utilities that have or plan atomic facilities would be wise not only to question management closely regarding the time periods and conditions for "yellowcake" contracts as well as the contracts for each stage of the fuel cycle, but also to read the actual contracts themselves, and then evaluate the probability of fulfillment as specified.

Since a utility board does not have to choose the "cheapest" power source either to meet competition or because of law, but is expected to make "prudent" decisions, why utility executives would risk the future of their companies with nuclear power remains a mystery to the authors. We find it interesting and perhaps significant that in the *Newsweek* of September 8, 1975, American Electric Power advertised that "For the third year in a row we are America's most efficient electric utility in power generation. . . ." American Electric utilizes coal almost entirely. Perhaps they know something others don't.

In this book, we have not sought to reach any philosophical, social, or environmental conclusions regarding any energy source. These areas have been touched upon only insofar as they influence an economic assessment. Based upon thorough in-depth analysis, the conclusion that must be reached is that, from an economic standpoint alone, to rely upon nuclear fission as the primary source of our stationary energy supplies will constitute economic lunacy on a scale unparallelled in recorded history, and may lead to the economic Waterloo of the United States. There can always be hope, however, that the candle of rigorous analysis will light the way through the tunnel of irrationality to the dawn of economic realism.

NOTE

1. Morgan Huntington, *The United States Nuclear Power Industry, a Summary of the Energy Yield and Fuel Demand*: A private memorandum dated April 20, 1975, gives 350 billion kwh to 1975. INFO news release, Atomic Industrial Forum, March 19, 1975 provided a figure of 170 billion net kwh for 1975.

TABLE A.1

Forecast of U.S. Energy Consumption and Electric Generating Capacity

Case		1973	1975	1980	1985	1990	1995	2000
Total Energy (10^{15} Btu)	Low	75.56	77.0	86.1	96.2	107.9	120.8	135.3
	Moderate/low	75.56	77.8	89.7	104.8	122.6	145.4	174.3
	Moderate/high	75.56	77.8	89.7	104.8	122.6	145.4	174.3
	High	75.56	78.6	95.3	116.6	136.8	162.2	195.0
Electric energy (10^{15} Btu)	Low	19.8	20.4	26.7	35.7	44.0	55.0	68.8
	Moderate/low	19.8	20.5	27.3	36.4	46.6	60.3	77.7
	Moderate/high	19.8	20.7	27.8	37.4	48.2	61.3	84.3
	High	19.8	20.9	28.9	39.9	52.9	71.0	96.8
Kilowatt hours (10^{9} kwh)	Low	1,878	1,940	2,570	3,500	4,400	5,555	7,020
	Moderate/low	1,878	1,955	2,630	3,570	4,660	6,090	7.925
	Moderate/high	1,878	1,972	2,675	3,660	4,820	6,400	8,600
	High	1,878	1,990	2,780	3,905	5,290	7,170	9,880
Heat rate (Btu/kwh)	Low	10,540	10,500	10,400	10,200	10,000	9,900	9,800
	Moderate/low	10,540	10,500	10,400	10,200	10,000	9,900	9,800
	Moderate/high	10,540	10,500	10,400	10,200	10,000	9,900	9,800
	High	10,540	10,500	10,400	10,200	10,000	9,900	9,800
System capacity factor (end of year)	Low	49.2	45.0	48.5	50.9	51.2	51.5	51.7
	Moderate/low	49.2	45.0	48.5	50.9	51.2	51.5	51.7
	Moderate/high	49.2	45.0	48.5	50.9	51.2	51.5	51.7
	High	49.2	45.0	48.5	51.0	51.2	51.5	51.7

Total Electrical capacity (GWe)	Low	436.0	492.0	605.0	785.0	980.0	1,230.0	1,550.0
	Moderate/low	436.0	496.0	620.0	800.0	1,040.0	1,350.0	1,750.0
	Moderate/high	436.0	500.0	630.0	820.0	1,075.0	1,420.0	1,900.0
	High	436.0	505.0	655.0	875.0	1,180.0	1,590.0	2,180.0
Nuclear	Low	18.4	37.2	70.5	160.0	285.0	445.0	625.0
	Moderate/low	18.4	38.5	76.0	185.0	340.0	545.0	800.0
	Moderate/high	18.4	40.4	82.0	205.0	385.0	640.0	1,000.0
	High	18.4	43.3	92.0	245.0	470.0	790.0	1,250.0
Hydro/pumped storage	Low	61.3	64.3	72.5	86.0	99.0	114.0	125.0
	Moderate/low	61.3	65.3	74.5	90.0	107.0	127.0	150.0
	Moderate/high	61.3	65.3	74.5	90.0	107.0	127.0	150.0
	High	61.3	65.3	77.5	94.5	114.0	136.0	165.0
International combustion/Gas Turbine	Low	37.8	43.3	51.0	57.5	64.0	74.0	85.0
	Moderate/low	37.8	44.3	53.1	60.0	70.0	85.0	105.0
	Moderate/high	37.8	44.3	53.1	60.0	70.0	85.0	105.0
	High	37.8	44.3	55.1	63.5	75.0	93.0	115.0
Fossil	Low	318.5	347.2	411.0	486.5	547.0	627.0	745.0
	Moderate/low	318.5	347.9	416.4	470.0	533.0	593.0	695.0
	Moderate/high	318.5	350.0	420.4	465.0	513.0	568.0	645.0
	High	318.5	352.1	430.4	472.0	521.0	571.0	650.0

TABLE A.2

Electrical Generating Capacity, Generation, and Energy Consumption

Year	Trillion Btu	Heat Rate	TWH	Cap. Fac.	GWe
U.S. Low Case, March 1975					
All Plants	*	*	*		*
1973	19,800.	10,543.	1,878.	49.2	436.0
1975	20,400.	10,515.	1,940.	45.0	492.0
1980	26,700.	10,389.	2,570.	48.5	605.0
1985	35,700.	10,200.	3,500.	50.9	785.0
1990	44,000.	10,000.	4,400.	51.3	980.0
1995	55,000.	9,901.	5,555.	51.6	1,230.0
2000	68,800.	9,801.	7,020.	51.7	1,550.0
IC/GT Plants		*		*	*
1973	497.	12,500.	40.	12.0	37.8
1975	569.	12,500.	46.	12.0	43.3
1980	670.	12,500.	54.	12.0	51.0
1985	756.	12,500.	60.	12.0	57.5
1990	774.	12,000.	64.	11.5	64.0
1995	856.	12,000.	71.	11.0	74.0
2000	894.	12,000.	74.	10.0	85.0
Hydroelectric*Plants					*
1973	2,936.	10,501.	280.	52.1	61.3
1975	2,830.	10,468.	270.	48.0	64.3
1980	3,020.	10,344.	292.	46.0	72.5
1985	3,370.	10,160.	332.	44.0	86.0
1990	3,540.	9,970.	355.	40.9	99.0
1995	3,930.	9,874.	398.	39.9	114.0
2000	4,280.	9,777.	438.	40.0	125.0
Nuclear*Plants			*		*
1973	857.	10,713.	80.	49.6	18.4
1975	2,200.	10,577.	208.	63.8	37.2
1980	4,330.	10,409.	416.	67.4	70.5
1985	9,500.	10,293.	923.	65.9	160.0
1990	16,900.	10,120.	1,670.	66.9	285.0
1995	25,800.	9,992.	2,582.	66.2	445.0
2000	35,500.	9,897.	3,587.	65.5	625.0
All – IC/GT Plants					
1973	19,303.	10,501.	1,838.	52.7	398.2
1975	19,831.	10,468.	1,894.	48.2	448.7
1980	26,030.	10,344.	2,516.	51.9	554.0
1985	34,944.	10,160.	3,440.	54.0	727.5
1990	43,226.	9,970.	4,336.	54.0	916.0
1995	54,144.	9,874.	5,484.	54.2	1,156.0
2000	67,906.	9,777.	6,946.	54.1	1,465.0
Thermal Plants					
1973	16,367.	10,501.	1,559.	52.8	336.9
1975	17,001.	10,468.	1,624.	48.2	348.4
1980	23,010.	10,344.	2,224.	52.7	481.5
1985	31,574.	10,160.	3,108.	55.3	641.5
1990	39,686.	9,970.	3,980.	55.6	817.0
1995	50,214.	9,874.	5,086.	55.7	1,042.0
2000	63,626.	9,777.	6,508.	55.4	1,340.0
Fossil and Other Central Station Plants					
1973	15,510.	10,489.	1,479.	53.0	318.5
1975	14,801.	10,452.	1,416.	46.6	347.2
1980	18,680.	10,329.	1,808.	50.2	411.0
1985	22,074.	10,103.	2,185.	51.8	481.5
1990	22,786.	9,862.	2,310.	49.6	532.0
1995	24,414.	9,751.	2,504.	47.9	597.0
2000	28,126.	9,630.	2,921.	46.6	715.0

Year	Trillion Btu	Heat Rate	TWH	Cap. Fac.	GWe
U.S. Moderate Low Case, March 1975					
All Plants	*		*		*
1973	19,800.	10,543.	1,878.	49.2	436.0
1975	20,500.	10,486.	1,955.	45.0	496.0
1980	27,300.	10,380.	2,630.	48.4	620.0
1985	36,400.	10,196.	3,570.	50.9	800.0
1990	46,600.	10,000.	4,660.	51.2	1,040.0
1995	60,300.	9.901.	6,090.	51.5	1,350.0
2000	77,700.	9,804.	7,925.	51.7	1,750.0
IC/GT Plants		*		*	*
1973	497.	12,500.	40.	12.0	37.8
1975	582.	12,500.	47.	12.0	44.3
1980	698.	12,500.	56.	12.0	53.1
1985	788.	12,500.	63.	12.0	60.0
1990	846.	12,000.	71.	11.5	70.0
1995	983.	12,000.	82.	11.0	85.0
2000	1,104.	12,000.	92.	10.0	105.0
Hydroelectric*Plants					*
1973	2,936.	10,501.	280.	52.1	61.3
1975	2,870.	10,437.	275.	48.1	65.3
1980	3,100.	10,334.	300.	46.0	74.5
1985	3,530.	10,155.	348.	44.1	90.0
1990	3,830.	9,969.	384.	41.0	107.0
1995	4,380.	9,873.	444.	39.9	127.0
2000	5,130.	9,779.	525.	39.9	150.0
Nuclear*Plants			*		*
1973	857.	10,713.	80.	49.6	18.4
1975	2,310.	10,596.	218.	64.6	38.5
1980	4,500.	10,393.	433.	65.0	76.0
1985	11,200.	10,323.	1,085.	67.0	185.0
1990	20,000.	10,116.	1,977.	66.4	340.0
1995	31,700.	9,994.	3,172.	66.4	545.0
2000	45,500.	9.898.	4,597.	65.6	800.0
All – IC/GT Plants					
1973	19,303.	10,501.	1,838.	52.7	398.2
1975	19,918.	10,437.	1,908.	48.2	451.7
1980	26,602.	10,334.	2,574.	51.8	566.9
1985	35,612.	10,155.	3,507.	54.1	740.0
1990	45,754.	9,969.	4,589.	54.0	970.0
1995	59,317.	9,873.	6,008.	54.2	1,265.0
2000	76,596.	9,779.	7,833.	54.4	1,645.0
Thermal Plants					
1973	16,367.	10,501.	1,559.	52.8	336.9
1975	17,048.	10,437.	1,633.	48.3	386.4
1980	23,502.	10,334.	2,274.	52.7	492.4
1985	32,082.	10,155.	3,159.	55.5	650.0
1990	41,924.	9,929.	4,205.	55.6	863.0
1995	54,937.	9,873.	5,564.	55.8	1,138.0
2000	71,466.	9,779.	7,308.	55.8	1,495.0
Fossil and Other Central Station Plants					
1973	15,510.	10,489.	1,479.	53.0	318.5
1975	14,738.	10,412.	1,415.	46.4	347.9
1980	19,002.	10,321.	1,841.	50.5	416.4
1985	20,882.	10,067.	2,074.	50.9	465.0
1990	21,925.	9,839.	2,228.	48.6	523.0
1995	23,237.	9,713.	2,392.	46.1	593.0
2000	25,966.	9,577.	2,711.	44.5	695.0

(Continued)

113

Year	Trillion Btu	Heat Rate	TWH	Cap. Fac.	GWe
U.S. Moderate High Case, March 1975					
All Plants	*		*		*
1973	19,800.	10,543.	1,878.	49.2	436.0
1975	20,700.	10,497.	1,972.	45.0	500.0
1980	27,800.	10,393.	2,675.	48.5	630.0
1985	37,400.	10,219.	3,660.	51.0	820.0
1990	48,200.	10,000.	4,820.	51.2	1,075.0
1995	63,300.	9,891.	6,400.	51.5	1,420.0
2000	84,300.	9,892.	8,600.	51.7	1,900.0
IC/GT Plants		*		*	*
1973	497.	12,500.	40.	12.0	37.8
1975	582.	12,500.	47.	12.0	44.3
1980	698.	12,500.	56.	12.0	53.1
1985	788.	12,500.	63.	12.0	60.0
1990	846.	12,000.	71.	11.5	70.0
1995	983.	12,000.	82.	11.0	85.0
2000	1,104.	12,000.	92.	10.0	105.0
Hydroelectric*Plants					*
1973	2,936.	10,501.	280.	52.1	61.3
1975	2,870.	10,437.	275.	48.1	65.3
1980	3,100.	10,334.	300.	46.0	74.5
1985	3,530.	10,155.	348.	44.1	90.0
1990	3,830.	9,969.	384.	41.0	107.0
1995	4,380.	9,873.	444.	39.9	127.0
2000	5,130.	9,779.	525.	39.9	150.0
Nuclear*Plants			*		*
1973	857.	10,713.	80.	49.6	18.4
1975	2,430.	10,611.	229.	64.7	40.4
1980	5,050.	10,412.	485.	67.5	82.0
1985	12,700.	10,300.	1,233.	68.7	205.0
1990	23,100.	10.096.	2,288.	67.8	385.0
1995	38,100.	9,995.	3,812.	68.0	640.0
2000	58,200	9,898.	5,880.	67.1	1,000.0
All – IC/GT Plants					
1973	19,303.	10,501.	1,838.	52.7	398.2
1975	20,118.	10,449.	1,925.	48.2	455.7
1980	27,102.	10,348.	2,619.	51.8	576.9
1985	36,612.	10,179.	3,597.	54.0	760.0
1990	47,354.	9,970.	4,749.	53.9	1,005.0
1995	62,317.	9,863.	6,318.	54.0	1,335.0
2000	83,196.	9,779.	8,508.	54.1	1,795.0
Thermal Plants					
1973	16,367.	10,501.	1,559.	52.8	336.9
1975	17,248.	10,449.	1,651.	48.3	390.4
1980	24,002.	10,348.	2,320.	52.7	502.4
1985	33,082.	10,179.	3,250.	55.4	670.0
1990	43,524.	9,970.	4,365.	55.5	898.0
1995	57,937.	9,863.	5,874.	55.5	1,208.0
2000	78,066.	9,779.	7,983.	55.4	1,645.0
Fossil and Other Central Station Plants					
1973	15,510.	10,489.	1,479.	53.0	318.5
1975	14,818.	10,422.	1,422.	46.4	350.0
1980	18,952.	10,330.	1,835.	49.8	420.4
1985	20,382.	10,104.	2,017.	49.5	465.0
1990	20,424.	9,832.	2,077.	46.2	513.0
1995	19,837.	9,620.	2,062.	41.4	568.0
2000	19,866.	9,445.	2,103.	37.2	645.0

Year	Trillion Btu	Heat Rate	TWH	Cap. Fac.	GWe
U.S. High Case, March 1975					
All Plants	*		*		*
1973	19,800.	10,543.	1,878.	49.2	436.0
1975	20,900.	10,503.	1,990.	45.0	505.0
1980	28,900.	10,396.	2,780.	48.5	655.0
1985	39,900.	10,218.	3,905.	50.9	875.0
1990	52,900.	10,000.	5,290.	51.2	1,180.0
1995	71,000.	9,902.	7,170.	51.5	1,590.0
2000	96,800.	9,798.	9,880.	51.7	2,180.0
IC/GT Plants		*		*	*
1973	497.	12,500.	40.	12.0	37.8
1975	582.	12,500.	47.	12.0	44.3
1980	724.	12,500.	58.	12.0	55.1
1985	834.	12,500.	67.	12.0	63.5
1990	907.	12,000.	76.	11.5	75.0
1995	1,075.	12,000.	90.	11.0	93.0
2000	1,209.	12,000.	101.	10.0	115.0
Hydroelectric*Plants					*
1973	2,936.	10,501.	280.	52.1	61.3
1975	2,870.	10,455.	275.	48.0	65.3
1980	3,220.	10,351.	311.	45.8	77.5
1985	3,700.	10,178.	364.	43.9	94.5
1990	4,080.	9,971.	409.	41.0	114.0
1995	4,690.	9,876.	475.	39.9	126.0
2000	5,640.	9.775.	577.	39.9	165.0
Nuclear*Plants			*		*
1973	857.	10,713.	80.	49.6	18.4
1975	2,590.	10.615.	244.	64.3	43.3
1980	5,860.	10,409.	563.	69.9	92.0
1985	15,350.	10,302.	1,490.	69.4	245.0
1990	28,900.	10,000.	2,890.	70.2	470.0
1995	48,200.	9,897.	4,870.	70.4	790.0
2000	74,900.	9,815.	7,631.	69.7	1,250.0
All – IC/GT Plants					
1973	19,303.	10,501.	1,838.	52.7	398.2
1975	20,318.	10,455.	1,943.	48.2	460.7
1980	28,176.	10,351.	2,722.	51.8	599.9
1985	39,066.	10,178.	3,838.	54.0	811.5
1990	51,993.	9,971.	5,214.	53.9	1,105.0
1995	69,925.	9,876.	7,080.	54.0	1,497.0
2000	95,591.	9.775.	9,779.	54.1	2,065.0
Thermal Plants					
1973	16,367.	10,501.	1,559.	52.8	336.9
1975	17,488.	10,455.	1,669.	48.2	395.4
1980	24,956.	10,351.	2,411.	52.7	522.4
1985	35,366.	10,178.	3,475.	55.3	717.0
1990	47,913.	9,971.	4,805.	55.4	991.0
1995	65,235.	9,876.	6,605.	55.4	1,361.0
2000	89.951.	9,775.	9,202.	55.3	1,900.0
Fossil and Other Central Station Plants					
1973	15,510.	10,489.	1,479.	53.0	318.5
1975	14,858.	10,427.	1,425.	46.2	352.1
1980	19,096.	10,333.	1,848.	49.0	430.4
1985	20,016.	10,085.	1,985.	48.0	472.0
1990	19.013.	9,927.	1,915.	42.0	521.0
1995	17,035.	9,815.	1,735.	34.7	571.0
2000	15,051.	9,579.	1,571.	27.6	650.0

THIS COMPUTER PROGRAM IS FOR DETERMINING SOME IMPLICA-
TIONS OF A GIVEN SET OF ASSUMPTIONS INVOLVING VARIOUS
TYPES OF ELECTRICAL GENERATING PLANTS INSTALLED IN A SYS-
TEM WITH RESPECT TO THEIR GENERATING CAPACITIES, ELECTRI-
CAL GENERATION, ENERGY CONSUMED FOR ELECTRICITY GENERA-
TION, AND THEIR RELATED HEAT RATES AND CAPACITY FACTORS.
IT WAS WRITTEN BY ARTHA JEAN SNYDER, OFFICE OF PLANNING
AND ANALYSIS, ENERGY R & D ADMINISTRATION, GERMANTOWN,
MD.

NUMBERS IN COLUMNS HEADED BY ASTERISKS FORM THE BASIS
FROM WHICH ALL OTHER NUMBERS ARE CALCULATED

Trillion BTU: Energy consumed for generating electricity in trillions (10^{12})
of British Thermal Units.

Heat rate: The number of BTUs required to produce one kilowatt-hour
of electricity.

TWH: Electricity generation in billions (10^9) of kilowatt-hours.

$$\frac{\text{Trillion BTU}}{\text{Heat rate}} = \text{TWH}$$

Cap Fac: Capacity factor as a percent.

GWe.: Installed electrical generating capacity in millions of
kilowatts.

$$\frac{\text{TWH} (10^9)}{\text{Cap Fac} \times 8{,}760} = \text{GWe}(10^6)$$

Since there are 8,760 hours in a 365-day year, a one-kilowatt rated
plant would have a theoretical output of 8,760 kilowatt-hours of electricity if
it operated 100 percent of the time. If we therefore take "1985, all plants
U.S. low case" as an example, we would have the following calculations:

$$\frac{3{,}500 \times 10^9}{.509 \times 8{,}760} = 785 \times 10^6$$

The final column "GWe" is the generating capacity necessary to meet
the demand in the TWH column given the capacity factor assumed. If the true
capacity factor were lower, and the demand remained the same, then more
plants (or GWe.) would be required to meet the demand.

TABLE A.3

Forecast of Energy and Electricity in the United States
(per capita)

		1973	1975	1980	1985	1990	1995	2000
Total energy (millions Btu)	Low	358	356	378	400	430	463	499
	Moderate low	358	360	393	437	488	557	643
	Moderate High	358	360	393	437	488	557	643
	High	358	364	418	486	545	621	720
Energy for electricity (millions of Btu)	Low	93.7	94.4	117	149	175	211	254
	Moderate low	93.7	94.9	120	152	186	231	287
	Moderate high	93.7	95.8	122	156	192	243	311
	High	93.7	96.8	127	166	211	272	357
Electric energy (kilowatt-hours)	Low	8,900	8,980	11,270	14,600	17,500	21,300	25,900
	Moderate low	8,900	9,050	11,540	14,900	18,600	23,300	29,200
	Moderate high	8,900	9,130	11,730	15,200	19,200	24,500	31,700
	High	8,900	9,210	12,200	16,300	21,100	27,500	36,500
Population (millions)		211	216	228	240	251	261	271

TABLE A.4

Fuel Cycle Lead Times

	Quarter Years
U_3O_8 from mill through UF_6 conversion	1
Enrichment	1
Fuel fabrication	
First cores	2
Reloads	1
Storage at reactor	
First cores	3
Reloads	2
Discharge to reprocessing	2
Reprocessing to re-use	
Uranium	2
Plutonium	1

ERDA COMPUTATIONAL METHODOLOGY

One of the great difficulties of working with ERDA data is that they do not include the methodology for getting from one column of numbers or table to the next. The jump is merely made. This explanation is provided as an attempt to remove some of the miasma that may be clouding the picture for the reader.

The tables in Appendix A have been computed by ERDA.* If one starts with their basic raw data and accepts this as accurate, all of the subsequent derived data appears to follow accurately. There are three significant problems, however, in following their line of reasoning and their calculations through from beginning to end.

1. Assumptions are changed and data bases modified.
For example, the tables giving the known resources of uranium (U_3O_8) are in short tons (2,000 lbs./ton) whereas demand is given in metric tons (2,204 lbs./ton). The tables on electricity demand and percent to be supplied by nuclear plants use capacity factors by approximately 65 percent, while tables on uranium demand are predicated upon 70 to 75 percent.

---

*Including predecessor AEC.

Where it is found convenient to manually translate data bases into equiv-
alences, such as converting short tons of resources to metric tons, this was
performed. In other cases where the task would have been unwieldy using
a manual electronic calculator, and where it was felt that the inconsisten-
cies were not significant to the evaluation of the overall scenario of the
nuclear fuel cycle, the computational errors were ignored. Forecasting 10
and 25 years is a highly speculative activity to begin with and as long as
there are not too many cumulative errors, 10 percent is not significant.

2. The assumptions underlying certain data are often either not presented,
   or not explicitly stated.

3. Intermediate steps are often omitted in the derivation of data.

To remedy this situation and to enable other researchers to more fully
understand ERDA releases, the following information is presented:

Referencing the electric energy consumption forecast in Table A.1, this
must be converted into required generating capacity. To do this, assumptions are
made regarding the capacity factors of different types of plants, the efficiency of
plants in converting heat to electricity, and the peak load requirements of the
energy system.

As of 1973, the overall capacity factor of all electrical plants in the United
States was 49.2 percent. ERDA forecasts that a low capacity factor will exist for
the next few years, and that there will be some long-term improvement in the
system factors, reflecting the modest introduction of peak pricing schemes
and/or more effective use of energy storage devices. Regarding capacity factors
for nuclear plants, ERDA has made the following assumptions:

- Plants achieve a 40 percent plant factor during pre-opera-
tional period between fuel loading and commercial operation.
- Plant factor is 65 percent during first two years of commercial operation.
- For years three through 15 of a plant, steady capacity
factors are assumed as follows:

High case	75 percent
Moderate/high case	72 percent
Moderate/low case	70 percent
Low case	70 percent

- Following the 15th year, plant factors decline at two percentage points per year to a minimum plant factor of 40 percent.*

---

*ERDA, *Total Energy, Electric Energy, and Nuclear Power Projections*, U.S., p. 6.

The projections show nuclear capacity factors escalating considerably from 49.6 percent in 1973 to 65.5 percent in 2000. This is seriously open to question. As pointed out in Chapter 5 on capacity factors, neither current nor past experience with either nuclear or fossil plants would lead to this high a capacity factor as a reliable assumption for a national average. Obviously, an overgenerous capacity factor will tend to understate the number of plants required.

Since all electrical generating plants are essentially machines for converting heat into electricity, the efficiency with which this is accomplished has a bearing upon the fuel consumed, the output of a plant, and the quantity of plants needed to meet a given demand level. The current U.S. average conversion rate is 10,500 Btu/kwh. Since the advent of central generating plants in the last century, the efficiency of plants has continued to improve. However, the rate of improvement is slowing down. By 1990, ERDA forecasts an average heat rate of 10,000 Btu/kwh, and by the year 2000, it should decline to 9,800. They are predicting that the heat rate of HTGRs (high temperature gas reactor nuclear plants) and FBRs (fast breeder reactors) will be 40 percent compared with 33 percent for current nuclear plants.

NOMOGRAPHS FOR NORMAL URANIUM FEED
AND SEPARATIVE WORK REQUIREMENTS

This appendix presents nomographs which may be used to determine the amounts of normal feed and separative work required to produce a unit of enriched uranium at any product assay up to 4.0 percent U-235, and at any diffusion plant tails assay from 0.1 percent to 0.4 percent U-235. The nomographs are followed by illustrations of their use.

The nomographs are intended for use in obtaining quickly a reasonably accurate estimate of feed and separative work requirements. They are neither sufficiently extensive nor precise enough for use where a full range of exact values is required.

NOMOGRAPH 1

Normal Uranium Feed Requirement

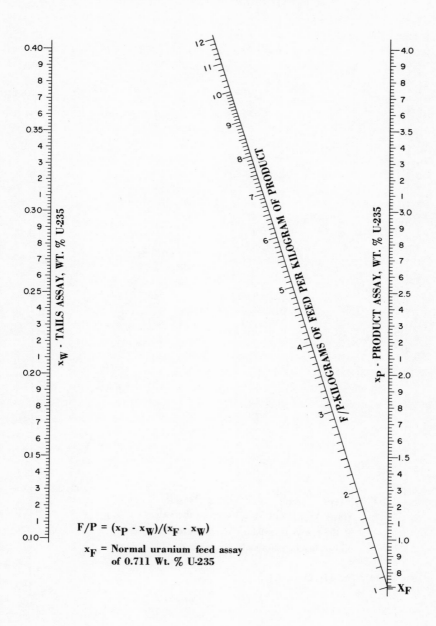

$$F/P = (x_P - x_W)/(x_F - x_W)$$

x_F = Normal uranium feed assay of 0.711 Wt. % U-235

NOMOGRAPH 2

Separative Work Requirement

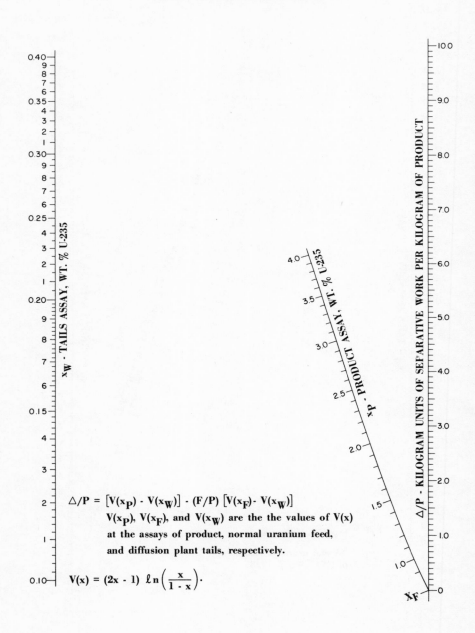

$\triangle/P = \left[V(x_P) - V(x_W)\right] - (F/P)\left[V(x_F) - V(x_W)\right]$

$V(x_P)$, $V(x_F)$, and $V(x_W)$ are the the values of $V(x)$ at the assays of product, normal uranium feed, and diffusion plant tails, respectively.

$V(x) = (2x - 1) \, \ell n\left(\dfrac{x}{1 - x}\right).$

ILLUSTRATED USES OF NOMOGRAPHS

Note: All quantities of uranium are as kilograms of contained uranium in uranium hexafluoride. All assays are as weight percent U-235.

NOMOGRAPH 1
Normal Uranium Feed Requirement

To find the amount of normal uranium feed required per kilogram of enriched uranium product, connect with a straightedge the tails assay of interest on the left line with the product assay of interest on the right line. The kilograms of normal uranium feed required per kilogram of enriched uranium product are found on the center oblique line at the point where it intersects the straightedge. To illustrate, a straightedge connecting a tails assay of 0.2% with a product assay of 2.5% intersects the feed requirement line at 4.50, the kilograms of normal feed required per kilogram of product under these conditions.

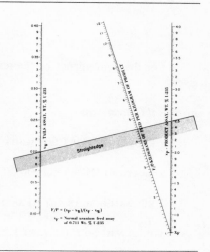

NOMOGRAPH 2
Separative Work Requirement

To find the amount of separative work required per kilogram of enriched uranium product when normal uranium feed is being enriched, connect with a straightedge the tails assay of interest on the left line with the product assay of interest on the center oblique line. The kilogram units of separative work required per kilogram of enriched uranium product are found on the right line at the point where it intersects the extended straightedge. To illustrate, a straightedge connecting a tails assay of 0.2% with a product assay of 2.5% intersects the separative work requirement line at 3.23, the kilogram units of separative work required per kilogram of product under these conditions.

THE MISSING LINKS

For the convenience of the reader the following numerical conversions are given:

Electrical Conversions

1 kilowatt (kw) = 1,000 watts

1 megawatt (MWe) = 1,000,000 watts

1 kilowatt-hour (kwh) = 1,000 (10^3) for one hour of time

1 megawatt-hour (MWe-hr.) = 1,000,000 (10^6) watts for one hour of time

So you see that a "kilowatt" refers to capacity or plant size, and a "kilowatt-hour" refers to output from that plant. Also:

1 MWe or 10^6 watts = 1,000 kw, or 10^3 kw

1,000 MWe or 10^9 watts = 1,000,000 kw or 10^6 kw

Do you see what has happened? When using scientific notation, we can subtract powers or exponents from one another. Since a kw = 10^3 watts, to convert 1,000 MWe or 10^9 watts to kw, we subtract 10^3, which is the conversion for watts into kilowatts, i.e.

1,000 megawatts   kilowatts
(10^9 watts)   – (10^3 watts) = 10^6 kilowatts

A mill is 1,000th of a dollar or one-tenth of a cent.

INTERVIEWS REGARDING COAL

Amax Coal Company, Indianapolis Office, March 13, 1975.

Alexander, Lloyd. Office of Coal, Federal Energy Administration, March 1975.

Beskerna, Charles. Vice President, Sales, Peabody Coal Company, March 1975.

Bressman, Bob. Rate Department, Chicago and Northwestern Railroad, March 18, 1975.

Budd. Wyoming Mining Association, March 13, 1975.

Fragmant, John. Coal Sales Division, Demmerer Coal Co., March 13, 1975.

Gakner, Alex. Economist, Federal Power Commission, March 1975.

Gibson, Otis. President, Illinois Coal Operators, March 13, 1975.

Hardcastle. National Coal Association, March 1975.

Hurdle, Kenneth H. Chief Statistician, Association of American Railroads, March 13 and August, 1975.

Linse, Norm. Market Manager of Energy Resources, Union Pacific Railroad, March 11, 1975.

Louks, Dr. Bert. Manager of Economics and Planning, Fossil Fuel Department, Electrical Power Research Institute, March 18, 1975.

Maxwell, Michael A. Chief, Non-Regenerable Processes Section, Control Systems

McCleod, Don. Vice President, Iowa Southern Utilities Company, February 3, 1975.

Laboratory, U.S. Environmental Protection Agency, April 1975.

Piper, W. G. Manager, Large Machine Sales, Bucyrus-Erie Company, March 13, 1975.

Skalla, Dean. Mining Division, Peter Kiewit & Sons, Inc., March 1975.

Zaffarano, Dr. Daniel. Director, Iowa Coal Research Project, January 31, 1975.

TABLE C.1

Results of Survey of Coal Operators (March 1975)

Region	Company or Association	Reserves Available for Contract?	Terms	Estimated Base Price — Strip	Estimated Base Price — Deep	Estimated Maximum Escalation
Wyoming	Amax	Yes	Based on ROI; long-term with base price, escalation clauses; most have 5-year renegotiation	$6.75/ton	–	10 percent and under
Wyoming	Kemmerer Coal Co.	Yes—but waiting strip mining regulations	Long-term; base price plus escalators	$8/ton	$15/ton	10 percent
Wyoming & Montana	Peter Kiewit & Sons, Inc.	Yes—but "waiting for strip mine regulations to be cleared up"	Long-term; base price plus escalators	$7-9/ton (9,500 Btu/lb. coal)	–	No more than 10 percent
Wyoming	Wyoming Mining Association	Not applicable. Said all companies "waiting to see what happens" with strip-mining regulations before signing contracts	–	–	–	–
Illinois	American Metal Climax (AMAX)	Yes	Based on ROI; base price with 14 or 15 terms escalated; long-term contracts, most with 5-year renegotiation clause	$15/ton	–	10 percent or under
Illinois	Peabody Coal Company	Yes—but currently tied up with antitrust suit; accepting no new contracts until settled	Long-term, base price with cost escalation clauses	$17/ton	$20/ton	10 percent or under
Illinois	Illinois Coal Producer's Association	Not applicable	Long-term (15 years or more) contract needed; base price plus escalation clause tied to cost factors; renegotiable periods getting more common	–	$20/ton	10 percent or under

126

CALCULATING OVERNIGHT PLANT COSTS AND CONSTRUCTION FACTORS

In a telephone conversation of July 30, 1975, F. C. Olds* gave the following information on 13 nuclear plants that were scheduled for completion before the end of 1975:

TABLE D.1

Survey of Nuclear Plant Estimated Completion Costs

Survey Date	Number of Plants	Average Size (MWe)	Estimated Cost at Completion in 1975
12/31/74	13	896	$382/kw
3/31/75	13	879	$401/kw
6/30/75	12	858	$416/kw

Of the 13 plants, three nuclear steam systems were ordered in 1968, nine in 1967, and one in 1966.

Cost escalation averaged 8.9 percent for the past six months or 17.8 percent per annum. The costs of building a plant are anticipated to be spread over about a nine-year period for nuclear plants beginning at this time. If expenditures were spread evenly over the construction period, a simple average of the estimates at the beginning of construction and the final cost would give an accurate average for the time period involved. Unfortunately, however, expenditures are made unevenly and the final cost is a weighted cost. But this difficulty can be overcome if the annual expenditures together with an inflation table are set in an array. Then the year in which the "average" cost is reached can be determined. In Table D.2, the percent shown spent each year are from breakdowns utilized by one Iowa utility in its planning. While there will be variations for each company, these appear to be representative. The annual percentages spent are:

---

*Senior Editor, *Power Engineering.*

Nuclear (over a nine-year period): 0.5, 1.0, 2.0, 6.5, 10.0, 40.0, 27.0, 10.0, 3.0

Coal (over a seven-year period): 1.5, 1.5, 8.0, 30.0, 46.0, 10.0, 3.0

The last figure in Column E will be the "construction factor" for a specific inflation rate as shown in Table 7.1.

TABLE D.2

Calculation Factors and Midpoint of Funds Expenditure for a Nuclear Plant

A	B	C	D	E	F*
Year	Percent Spent	15 Percent Compounded	B × C	Σ D	Cumulative Percent of Total Spent
1	0.5	1.150	.00575	.00575	0.238
2	1.0	1.322	.01322	.01897	0.784
3	2.0	1.521	.03042	.04939	2.041
4	6.5	1.749	.11369	.16308	6.742
5	10.0	2.011	.20110	.36418	15.054
6	40.0	2.313	.92520	1.28938	53.302
7	27.0	2.660	.71820	2,00758	82.991
8	10.0	3.059	.30540	2.31348	95.637
9	3.0	3.518	.10554	2.41902	100.000

* Number in Col. D ÷ 2.41902

Source: Compiled by the authors.

From this table, it can be seen that the midpoint for the total funds expended takes place in year "6" and that there are three more years to completion of the project. After the "average" is reached, there is still 52.1 percent more on the balance of the project. By adding 52.1 percent to the $416 "average" completion cost, an "overnight" cost of $633 is obtained. This approximates Olds's estimate. (While the data in Table D.2 is based on a nine-year construction period, and the data in Table D.1 implies an eight-year period, funds are typically expended before ordering the nuclear steam system.)

METHODOLOGY FOR COMPUTING CAPACITY FACTORS

An independent analysis of coal-fired plant capacity factors was performed because the available data on fossil fuel capacity factors was so unspecific that there was no adequate basis for comparison with the studies of nuclear plant capacity factors such as those published by David Comey. Previous to this report, the best available data on fossil fuel capacity factors came from the Edison Electric Institute, as quoted in the text. However, the EEI compilation had certain shortcomings:

- It did not distinguish between coal-fired and other fossil fuel plants.
- It did not list fossil-fuel capacity factors by the age of the plants.
- It contained data from very old as well as very new plants, whereas all nuclear plants are relatively new.
- The size category compilations offered were not comparable to Comey's qualifications of a minimum 100 MWe size for nuclear plants.

Thus, in order to provide an "apples and apples" comparison between Comey's data and the Federal Power Commission and National Coal Association reports (for 1973 data) on electric utility capacity factors that were surveyed, the following criteria were established for inclusion within the survey:

- The plant must be coal-fired.
- It must have been at least 100 MWe in size in its initial entry in the FPC reports.
- It must have been initially reported in one of the FPC reports, *Steam-Electric and Plant Construction Cost and Annual Expenses*, Annual Supplement Numbers 14 to 25 (1961-72) in order that the plants would be in the same age category as most nuclear plants.
- It must have had at least one full year of operation (interpreted as a recorded capacity factor reading) before the 1972 supplement.

A total of 68 coal-fired plants met these age and size criteria. After screening these publications for plants that fit within the framework of the criteria established, the rated capacity and capacity factor for each year was recorded. Each "plant/year" of data then became a "unit." Utilizing these "units," three analyses were performed:

- An overall average of all "units" was calculated, weighted by size of plant.
- Three size categories were established: 1) Under 500 MWe, 2) 500 to 1,000 MWe, and 3) over 1,000 MWe. Each "unit" was then placed in the appropriate size classification and the capacity factor of each size group was calculated, weighted by size of plant.
- In the third analysis of plants by age, certain "units" were first eliminated. Since it is common for a utility to increase the capacity of an individual plant, a hybrid of old and new equipment results, thus making future analysis by age not really valid. With these hybrids omitted, all other plant/year "units" were then averaged by age of plant.

SOLAR POWER

Because solar power will present the consumer with at least a partial alternative to the purchase of energy from a central distribution system, the sale of electrical units by power companies may begin to decline from the future levels currently projected; as discussed in Chapter 10, this may seriously impact the financial stability of utilities. Therefore this appendix will explore the potential economics of solar cells during the next decade, when they may begin to offer competition to conventional power sources. But, this appendix will not exhaustively examine all types of solar systems nor rigorously define its economics. Because solar energy is only now on the threshold of becoming economically viable as an energy alternative, the costs shown are solely estimates based on today's best available data.

We have not intended to imply that other energy sources such as geothermal or wind power do not have promise, but solar energy is being more broadly researched and would appear to have more universal geographic application. Thus, because of its potential immediacy, the promise, problems, and possible costs of solar power do require examination. Solar power is a renewable energy source; it is never depleted and has few environmental problems. Two basic approaches are available for harnessing the sun's energy:

- flat plate collectors or other devices that collect the sun's heat and then heat or cool buildings
- direct generation of electricity

The technology for applying these methods is now available, but two problems hinder the expansion of solar applications. They are:

- the high cost of what are essentially custom-built units
- the lack of effective energy storage devices that are also economical

STAYING COMFORTABLE WITH THE BLACK BOX

The devices marketed today for heating usually employ a "black box" flat plate collector. In this design, a black coated or other heat absorbing surface such as aluminum is covered with one or two sheets of glass to trap the heat. Air or water is then circulated through the box and transferred to an energy storage system made of hot rocks, eutectic salts, or water-filled tanks. Then, as required, the heat is circulated through the building.

The amount of space occupied by a unit of this type poses no problem since the roof area of a typical home usually contains more than enough square footage. The major barrier to market demand for these types of units is the $4,000 to $5,000 cost for a device that provides between 40 percent and 80 percent of the homeowner's needs. At today's fuel prices, the savings in fuel are insufficient to warrant spending this amount of money. If fuel prices rise sufficiently, or if the cost of a unit can be cut in half through mass production techniques, a unit would then pay for itself within five years and become an attractive investment.

FIGURE F.1

Solar Heated House

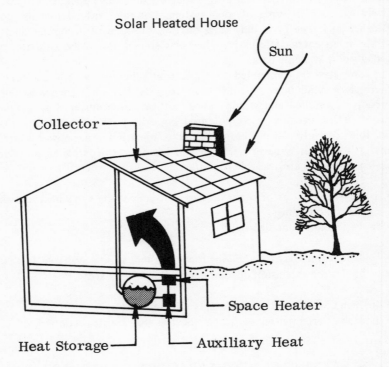

When a solar black box unit is used for cooling, the principle of the old gas refrigerator is employed. Since no standard unit of this type is being manufactured today, a cost range is not available. In 1968, space and water heating for the residential and commercial sectors accounted for 21.9 percent of total U.S. energy consumption, while air conditioning was responsible for 2.5 percent. Thus, the eventual saturation point for this type of technology would be 20 to 25 percent of U.S. energy consumption.[1]

ELECTRICITY FROM THE SUN

A technology with the promise of more widespread use is that of solar cells, which are small wafers of electrically conductive material that generate direct current when exposed to sunlight. These units offer the promise of being:

- pollution-free
- silent
- often without moving parts, thus making them highly reliable and cheap to maintain
- inflation-proof since after installation, there are no rising fuel costs to contend with
- long-lived
- modular, which means they can be placed at the location of the user, thus obviating an extensive distribution network
- made in a factory that would have no more than a one year lead time from the start of plant construction to the beginning of production of units immediately available for use (This creates a foundation for unheard-of flexibility in energy planning and contrasts with the 10-year lead time for nuclear plants.)

The most commonly used material for solar cells is silicon, the earth's second most abundant element, found in sandy beaches and deserts everywhere. Silicon is a semiconductor material, which means that it is simultaneously an electrical conductor and an insulator. Its best known use at this time is as the "guts" of pocket calculators.

To date, most applications of solar cells have either been for the space program or for military use. Labor intensive batch processing is used, beginning with the "growing" of silicon crystals in a furnace with a precisely controlled chemical and temperature atmosphere. These long crystals, which look like somewhat symmetrical carrots or bananas, are then sliced with diamond saws into thin wafers. This step alone wastes at least a third of the crystal. Next, the wafers are coated with boron to produce a positive electrical layer that interacts with the negatively charged silicon. The wafer is then cut into squares two centimeters on a side and placed in a rigid carrier; finally, six tiny silver wire connectors are imbedded in the cell. The process takes days and requires the use of highly skilled labor.

At the present time, solar cells for terrestrial use sell for $20,000 per peak kilowatt[2] compared with $640/kw for a nuclear plant—a factor of 31X. That a cost reduction of this magnitude can be achieved has been demonstrated twice in the last two decades with similar technologies. First, the cost of transistors was reduced by a magnitude greater than 31, and the same was accomplished with semiconductor electronic circuits.

THE POTENTIAL OF MASS PRODUCTION

Two approaches to bringing down the costs of solar cells are now in the forefront. The first of these is the "edge-defined, film-fed growth" (EFG) technique being developed by Mobil Tyco Solar Energy Corporation of Waltham, Massachusetts, a firm 80 percent owned by Mobil Oil and 20 percent by Tyco labs. Instead of the batch processing method, purified silicon is fed into a heated crucible, melted, and drawn through a die in a continuous "ribbon." To date, ribbons up to 50 feet long with a 10 percent efficiency rate have been successfully manufactured. This technique is not new to Tyco; it has been used for years in the production of sapphire ribbons for use in high pressure sodium vapor lamps.

An alternate approach is being taken by Varian Associates of Palo Alto, California. They have developed an aluminum gallium arsenide (AlGaAs) cell, which has a 21 percent to 23 efficiency rate compared with 10 percent for silicon. Because the AlGaAs can stand much higher temperatures than silicon, Varian is able to use concentrators such as mirrors or lenses, which increase the concentration of sunlight on the cell by as much as 1,000 times.

Using a parabolic mirror that concentrated direct sunlight 896 times, experimental cells demonstrated an efficiency of 17.2 percent and an output power density of 130 kw per square meter. Ronald L. Bell, director of Varian's solid-state laboratory, maintains that efficiencies as great as 23 percent can be achieved with other simple concentrator schemes, and that a 10 MWe peak-output plant would require only 80 square meters of gallium arsenide cells compared with 24.5 acres of cells for an equivalent system using 12 percent efficient silicon cells.[3]

On June 7, 1976, the *Wall Street Journal* reported that a team of scientists and engineers at the Massachusetts Institute of Technology, performing research sponsored by National Patent Development Corp. of New York, had developed a solar energy unit that increased the sunlight converted by a factor of between 500 and 1,000. The device is expected to be on the market by the end of 1977 and would cost about $5,000 for a single-family home. This unit would produce half of the electricity and about 70 percent of the hot water, heat and air conditioning for a house, according to Jerome I. Feldman, president of National Patent Development.

ENERGY STORAGE: A REQUIRED ACCESSORY

The sun obviously does not shine all of the time, and even when it does, its strength is often reduced by atmospheric conditions. Therefore, storage devices become necessary elements of a solar power system. Research is being conducted simultaneously on a number of approaches to solve this problem economically.

These include, but are not limited to:

- batteries: lead-acid, zinc-chlorine, sodium-sulfur, lithium-iron, and other advanced types
- hydrogen storage
- kinetic energy storage through the use of mechanical flywheels
- hydro pumped water storage
- compressed air storage
- steam storage
- superconducting magentics, and
- thermal energy storage

According to F. R. Kalhammer and P. S. Zygielbaum of the Electric Power Research Institute, most systems except lead-acid batteries are three to five years away from commercial development.[4] It would appear that with greater research funding, this time period could be shortened.

EFFECTIVE OUTPUT FROM SOLAR CELL SYSTEMS

The computation of solar cell output begins with the computation of sunlight available per given area. The quantity of sun energy falling on Washington, D.C. at noon on a clear day in June is often chosen as the starting point for calculations. Under these optimum conditions, 10 kw of electricity falls on 10 square yards of land. With cells having a 10 percent efficiency, 1 "peak kw" can be generated.[5]

Since the sun naturally is not shining at night, and since inclement weather and other atmospheric conditions such as fog, smog, and clouds, plus seasonal and hourly differences in light intensity reduce the quantity of sunlight, the expected availability of sun is an average of 18 percent of peak power for the United States as a whole and 25 percent for the Southwest.[6] Thus, 10 square yards of 10 percent efficient solar cells would produce 1,576 kwh/yr. throughout the United States and 2,190 kwh/yr. in the Southwest.*

Further reductions in the net energy available will occur because of the need for storage systems that will have an energy loss between 10 percent and 40 percent.[7] If it is assumed that there is a 35 percent loss in a storage system and that 75 percent of the power used must be stored, the net energy computation is:

---

*8,760 kwh/yr. × 18% = 1,576 kwh
8,760 kwh/yr. × 25% = 2,190 kwh

Average U.S.

      1,576 kwh × 25%   =394 kwh

      1,576 kwh × 75% × 65% =768

                              1,162 kwh/yr.

Southwest

      2,190 kwh × 25%   =   578

      2,190 × 75% × 65%   =1,068

                              1,646 kwh/yr.

CAPITAL COSTS

There are four main components within the capital cost framework of a solar installation: 1) costs of cells, 2) mounting structures, 3) storage systems, and 4) inverters. For a central power system, the cost of land and the distribution system must also be added. The current cost of purified silicon is $30/lb., and it is anticipated that this will drop to $15/lb. as demand increases.[8] At $30/lb., the estimated cost per peak kilowatt of power from a system using silicon solar cells of the Tyco design would be:[9]

Prime costs:	Silicon	$150
	Manufacture of ribbon	75
	Conversion to solar cells	75
	Packaging	50
		$350
Overhead and profit		$350
		$700/peak kw

At a silicon cost of $15/lb., this would be reduced to $625/peak kw. These are the projected costs for an array of 10 square yards. Dr. Mlavsky of Tyco is confident that these price objectives can be achieved and that Tyco can be ready for mass production between 1972 and 1985.[10]

If the support structures for an array were $2.00/sq. ft., the cost for a 10-square yard array would be $180/peak kw. A possible cost for the energy storage system would be $240/peak kw,[11] and for a DC-to-AC inverter, $60/peak kw.[12] The total basic cost for a peak kilowatt would then be:

Solar cells	$700
Structure	180
Storage system	240
Inverter	60

$1,180/peak kw

If this were mounted on the roof of a building or over a parking lot or other multipurpose site, there would be no additional expenses. At a 10 percent interest cost and a 30-year expected life, the annual cost would be $125.17.* With a projected output of 1,162 kwh/yr., the cost per kwh would be 10.8¢. This compares with a present power company delivered price of 3.5 cents, which is typical. If all power today were to be delivered from new plants rather than from a mix of old, cheaper plants and current expensive ones, the current delivered price to consumers would be about 7.0 cents. Over the next 10 years, as new plants constitute an increasingly larger part of the rate base, rates will begin to approach the higher dominant cost.

If the average electric bill were to rise for the next 10 years at a 12 percent annum rate (which takes into account not only general inflation, but also the increasing proportion of "net plant" accounted for by new additions), by 1985 a utility company 10.9¢/kwh rate will make solar cells competitive if they can be produced at the prices discussed earlier.

A central station solar cell unit would incur the additional expenses of land and distribution, but the cost of land would not be significant. Even if $2,000 per acre were paid—the price of high quality Iowa farmland—with a potential of locating 484 peak kilowatts per acre, the cost would be only $4.13/kw. On the other hand, the costs of distribution, plus overhead, plus profit would boost the cost of delivering power by 50 percent if the current approximate 40/60 ratio between cost of generation and all other costs is maintained. At a 15 percent capital recovery rate, the cost of delivered power would be 38.7¢/kwh.† Because of the necessity for a utility to earn a rate of return higher than the cost of mortgage money available to an individual, and because of the need for a distribution system not required by a homeowner, it would appear that the Tyco type of silicon cells would not be appropriate to central station power generation, at least initially.

By contrast, the Varian GaAs system, which utilizes a much more expensive unit as its base but which may have a lower kilowatt cost than the Tyco cell,

---

*A 10 percent interest charge is the cost of a homeowner mortgage compared with the overall 15 percent capital recovery cost of a utility.

†

$$\frac{0.1523\,[1,180 + (1.5)\,(1,180)]}{1,162}$$

may lend itself more readily to central station use. Assuming that the cost of a GaAs cell were high, say $20,000/peak kw, a concentration of 900 times the sunlight on such a cell would bring costs down to $23/peak kw for the cell component. This leaves plenty of margin for an elaborate concentrating system of mirrors and lenses. And while the per kilowatt cost may be low, the output would exceed the needs of a home and the overall cost would be too high for an individual. One more alternative is the "neighborhood" unit, which would service a half dozen or so houses with shared costs. Only time will tell what approach will become dominant.

LAND AVAILABILITY AND REQUIREMENTS FOR SOLAR CELLS

Assuming that solar cell units do become cost effective for central station systems, the question of land needs becomes relevant. Under a National Science Foundation contract, the Aerospace Corporation performed an analysis of the southwestern United States, utilizing the most stringent set of criteria (which excluded all land suitable for agriculture and mineral exploration, all federal lands and Indian reservations, all vegetated areas of any significance, and all areas unsuitable for siting because of terrain, erosion, or unsettled soil). Within these parameters, Aerospace identified over 21,000 square miles of land in the southwestern part of the country, which would conceivably lend itself to the erection of solar collectors.[13]

To ascertain what the maximum potential land requirements might be, we can assume that all nuclear power projected to enter service between 1985 and 2000 is provided by solar cells instead. From Appendix A, the data in Table F.1 can be extracted. In Table F.2, the figure of 18,067 square miles, which would be utilizing essentially worthless land, still comprises only 0.5 percent of the total area of the United States. Even without utilizing the tops of existing structures and other multi-purpose land, it becomes evident that insufficient land is not a barrier to the use of solar cells.

TABLE F.1

Incremental Nuclear Generation, 1985-2000 (Billions of kwh annually)

Low Case	Moderate/Low	Moderate/High	High
2,664	3,512	4,647	6,141

TABLE F.2

Square Miles Required for Solar Cell Systems in Year 2000

Low Case	Moderate/Low	Moderate/High	High
7,837	10,332	13,672	18,067

The calculation of land area required is:

In southwest U.S., 1,646 kwh/yr. are produced by one 10 sq. yd. array. Add 50% for access roads, pathways, structures, etc.

$$\frac{640 \text{ acres/sq. mi.} \times 4,840 \text{ sq. yds./acre}}{15 \text{ sq. yds./array}} = 206,507 \text{ arrays/sq. mi.}$$

206,507 arrays/sq. mi. $\times$ 1,646 kwh/array = 3.399 $\times$ 10^8 kwh/sq. mi.

The "low case" calculation is:

$$\frac{2,664 \times 10^9 \text{ kwh}}{3.399 \times 10^8 \text{ kwh/sq. mi.}} = 7,837 \text{ sq. mi.}$$

SUMMARY

Solar heating unts are commercially available today, but they are not competitive with the cost of central station power. Various approaches to making solar energy competitive with conventional sources are being pursued by the government as well as private corporations. No technological breakthroughs are required, but engineering advances are necessary to refine the experimental models and to reduce production costs. Also, storage devices must be made available at cost effective prices.

One of the most promising developments is that of solar cells, which directly convert sunlight into electricity. Two corporations appear to be making significant progress, and even without government assistance, commercial production can be anticipated at least by 1982.

CONCLUSION

It can be seen from the data presented in this appendix that with reasonable assumptions about future solar cell costs under mass production, solar cells should be competitive with conventional power sources after they are introduced in the mid-1980s. As fuel costs for conventional sources continue to rise, solar power should begin to gain a competitive edge.

Considering that except for coal, the fuels of conventional power sources will begin to be seriously depleted by the late 1980s, greater reliance will be placed upon renewable energy sources. Whether solar power achieves the dominant position of which it is capable will be strongly influenced by government priorities, research, and capital assistance as well as the presence or absence of tax incentives. However, neglect of solar energy could delay its implementation, since further engineering is required to reduce production costs; moreover, substantial investments will be necessary to attain the mass production levels required. With the short lead times required to bring a manufacturing plant on line, solar cell systems can make a significant contribution to U.S. power needs beyond the next decade.

NOTES

1. Stanford Research Institute, 1972, using Bureau of Mines and other sources.

2. U.S. House of Representatives, *Hearings before the Subcommittee on Energy of the Committee on Science and Astronautics*, June 6th and 11th, 1974, p. 28, Testimony of Eugene L. Ralph, Vice President, Research and Development, Spectrolab, a division of Textron, Inc.

3. *Electronics*, "Solar Cell Offers 21% Efficiency with GaAs," May 29, 1975.

4. F. R. Kalhammer and P. S. Zygielbaum, *Potential for Large-Scale Energy Storage in Electric Utility Systems*, ASME Publication No. 74-WA 1 Ener-9, presented at the Winter Annual Meeting of the American Society of Mechanical Engineers, New York, November 17-22, 1974.

5. U.S. House of Representatives, op. cit., p. 3, Testimony of Dr. Joseph J. Loferski, Division of Engineering, Brown University.

6. The Mitre Corporation, *Systems Analysis of Solar Energy Programs*, MTR-6513, report under contract to National Science Foundation, December 1973, p. 91.

7. Kalhammer and Zygielbaum, op. cit., p. 11.

8. U.S. House of Representatives, op. cit., p. 55, Testimony of John V. Goldsmith, Group Supervisor, Solar Energy Group, Jet Propulsion Laboratory, California Institute of Technology.

9. Ibid., pp. 117-21, Testimony of Dr. A. I. Mlavsky, Senior Vice President, Tyco Laboratories.

10. Telephone conversations with Dr. A. I. Mlavsky, spring 1975.

11. Kalhammer and Zygielbaum, loc. cit. These EPRI researchers estimated capital costs for energy storage systems ranging from $30/kw for electromagnetic storage to $250/kw for a combination hydrogen electrolysis and fuel cell system.

12. Mitre Corporation, loc. cit. Figure of $60/kw is double Mitre estimate, which was in 1970 dollars.

13. Aerospace Corporation, *Solar Thermal Conversion Mission Analysis, Summary Report—Southwestern United States*, prepared for the National Science Foundation, research applied to national needs. January 1975, p. 49, Chart 20.

WASTE AS FUEL

Just as the breeder reactor is an extension of current nuclear technology and therefore required at least some discussion, so does "waste as fuel" technology, especially since it is used in conjunction with the burning of coal. Understandably, most people are skeptical about turning a "sow's ear into a silk purse," but the technology of using solid waste as fuel not only obviates the difficulty of disposing of growing mountains of waste, but economically serves as fuel by substituting for coal.

HOW THE PROCESS WORKS

Domestic solid waste gathered through ordinary collection procedures is shredded, and the pieces are separated into light and heavy segments by an air blower. The light segment is then blown directly into the boiler and burned along with the coal; the heavy, noncombustible segment is passed by a magnet that separates out ferrous metals. Some systems also recover aluminum and glass from the noncombustible material. Finally, what remains represents, by volume, 5 percent of the original waste and this is used as landfill.

SUCCESSFUL RESULTS

In April 1972, the Union Electric Company and the city of St. Louis began commercial operation of the first large-scale demonstration project to use prepared solid waste as a supplement in a coal-fired generating unit. Two indentical boilers at the company's Meramec plant, each with a nominal rating of 125 MWe, were equipped with solid waste firing equipment. Because Union Electric considers that the system has operated successfully, the utility has announced plans to utilize all of the municipal waste in the St. Louis metropolitan area, and 17 other communities, impressed with the results, have constructed or have announced plans to build similar facilities.[1]

Although the St. Louis system was originally designed, on a heat value basis, to replace 10 percent of the coal with waste, the system has been found to operate well with 15 percent substitution. Utility engineers say that even 20 percent is realistic for existing boilers and 25 percent to 30 percent is possible for new boilers specially designed to use solid waste materials.[2] Further testing is planned to determine the maximum percentage of substitutability that can be achieved.

SYSTEMS REQUIREMENTS AND COSTS

If a 1,000 MWe coal-fired plant were designed to use waste for 25 percent of its fuel requirements, the waste feed processing capacity of the plant would be 10,132 tons per day while the firing equipment would be required to handle 6,333 tons per day.*

The cost of shredding and separation processing equipment would be $3,750/ton of daily capacity and the firing facilities to inject the shredded waste into the coal boiler would be $3,375/ton.[3] Total cost for the system would be $61.4 million.† This system would process 2.3 million tons per year. At a 30-year 15 percent capital recovery factor, the capital cost would be $4.07/ton.‡

The operating expenses incurred in processing the feed material include mostly cost of labor, which EPA estimated at $4.50/ton of feed. Where the processing plant is adjacent to the power plant, the transportation cost was projected at $1.00/ton,[4] for a total operating expense of $5.50/ton. Adding 50

---

* Calculation Methodology

Electrical output from waste = 1,000 MWe $\times$ 24 hrs./day $\times$ 25% = 6,000 MWe/hrs./day

Plant conversion heat rate = 9,500 Btu/kwh or 9.5×10^6 Btu/MWe-hr.

Waste Btu requirements = (6,000 MWe hrs./day) (9.5×10^6 Btu/MWe-hr.) = 57×10^9 Btu/day

If the raw waste feed has a heat value of 4,500 Btu/lb or 9.0 million Btu/ton, then

$$\frac{57 \times 10^9 \text{ Btu/day}}{9 \times 10^6 \text{ Btu/ton of feed}} = 6,333 \text{ tons of feed/day}$$

Since garbage collection normally takes place on five days out of a seven-day week (subtract one day for occasional holidays), the processing capability must be sufficient for eight days of fuel.

$$\frac{6,333 \text{ tons/day} \times 8 \text{ days}}{5 \text{ days}} = 10,132 \text{ tons/day of processing capability}$$

$$
\begin{array}{ll}
^{†}10,132 \times \$3,750 & = \$40.0 \text{ million} \\
\phantom{^{†}}6,333 \times \$3,375 & = 21.4 \\
\hline
& \$61.4 \text{ million}
\end{array}
$$

$$\frac{^{‡}(61.4 \times 10^6)(0.1523)}{2.3 \times 10^6} = \$4.07/\text{ton}$$

percent for inflation gives a grand total of $8.25/ton. Total system costs would therefore be:

Capital	$ 4.07/ton
Operating costs	8.25
Total cost	$12.32/ton of raw solid waste feed

Now that the expenses are known, it is relevant to ascertain what the potential revenues are in order to arrive at a net economic cost.

INCOME AND SAVINGS

In the separation process, materials of value are segmented and subsequently sold. At St. Louis, ferrous metals are recovered and sold at a rate equal to approximately $1/ton of raw solid waste.[5] At Ames, Iowa, greater classification will take place, and the sale of glass, aluminum and potash in addition to ferrous metals is expected to yield $3/ton.[6] An average operation that recovered ferrous metals and some, but not all, other materials might be expected to generate revenue of $2/ton.

Savings in landfill costs is another economic benefit from the process. Depending upon the area of the country, the availability of land, transportation costs, and the environmental acceptability of the operation to local cities, landfill costs may range from $3 to $5/ton. Since the recovery process reduces waste to 5 percent of its original volume, a $4/ton savings on landfill operations would seem reasonable. The "profit and loss" statement would now read:

Gross cost of process	$12.32/ton
Income and savings	6.00
Net cost	$ 6.32/ton

COST COMPARISON WITH COAL

Utilizing the data from Table 3.6, we can compare the delivered costs of coal and waste as fuel using Des Moines, Iowa as an example.

TABLE G.1

Comparison of Solid Waste and Coal

	Heat Value	Cost
Waste	4,500 Btu/lb.	70.22¢/million Btu
Western coal (surface-mined)	10,500	57.15
Illinois coal (deep-mined)	12,560	88.97

From these numbers, it can be seen that at today's prices, waste can be competitive with coal. As fuel costs continue to rise with inflation, waste should become even more glamorous as a fuel source. For utilities remote from coal mines, this day should already have arrived.

THE NATIONAL POTENTIAL

In 1971, 125 million tons of municipal solid waste were collected from U.S. residential and commercial sources and discarded with no attempt to recover the energy contained.[7] Not all the waste available would be economically recoverable, but the 75 percent that comes from metropolitan areas with over 50,000 population would be feasible to use.[8] With a heat value of 9.0 million Btu/ton of raw waste,[9] this represents 844 trillion usable Btu. With an average of 24.8 million Btu/ton of coal,[10] 34 million tons of coal could have been saved, an amount equal to over 10 percent of the coal mined in the United States for electrical purposes that year.[11]

SUMMARY

Solid waste has been successfully used on a commercial scale in coal-burning plants. If Des Moines, Iowa is cited as a typical location, the cost in terms of "cents per million Btu" is 70.22 cents for waste, which compared with 57.15 cents for western strip-mined coal and 88.97 cents for Illinois deep-mined coal. The available and usable waste from metropolitan areas is so great as a potential energy source that approximately 10 percent of the coal used for electricity could be saved.

CONCLUSION

Energy from solid waste appears to be economically viable now, and with further development can become a material aid in meeting national energy goals.

NOTES

1. EPA, Office of Solid Waste Management Programs, *Energy Conservation Through Improved Solid Waste Management*, EPA-SWA-125 (1974): II. 11-15. The cities or districts are: Bridgeport, Connecticut; District of Columbia; Chicago; Chicago area (excluding city); Ames, Iowa; Montgomery County, Maryland; East Bridgewater, Massachusetts; Essex County, New Jersey; Albany, New York area; Monroe County, New York; New York City; Lane County, Oregon; Philadelphia, Pennsylvania; San Juan, Puerto Rico; Memphis, Tennessee.

2. Interview with Dave Klump, Project Manager, Solid Waste Utilization Project, Union Electric Company, January 1975.

3. EPA, *Energy Recovery from Waste*, EPA-SW-36d (1973): 18. EPA figures for 1973 have been escalated by 50 percent.

4. Ibid.

5. Ibid.

6. Interview with Jerry Temple, Solid Waste System Superintendent, Ames, Iowa Power Plant, January 1975.

7. EPA-SWA-125, op. cit., p. 5.

8. Ibid.

9. EPA, Office of Solid Waste Management Programs, *Markets and Technology for Recovering Energy from Solid Waste*, EPA/530/SW-130 (1974): 1.

10. National Coal Association, *Bituminous Coal Facts, 1972*, p. 85.

11. Ibid., p. 64.

Aerospace Corporation. *Solar Thermal Conversion Mission Analysis Summary Report–Southwestern United States* Volume 1, prepared for the National Science Foundation, research applied to national needs, January 1975.

American Association of Railroads. *Statistics of Railroads of Class 1 1963-1973.* August 1974.

Atlanta Journal and Constitution Magazine, "The Most Hazardous Business," February 2, 1975.

Baranowski, Frank P. *Uranium Supplies, Costs and AEC Resource Evaluation Program.* Remarks given to the General Advisory Committee to the USAEC, 130th meeting, November 7, 1974.

Bureau of National Affairs, Bulletin No. 72, December 26, 1974.

Bureau of National Affairs, Bulletin No. 91, May 8, 1975.

Carter, Gov. James. Letter to Robert M. Lazo, Esq., Chairman, Atomic Safety and Licensing Board Panel, U.S. Atomic Energy Commission, January 6, 1975.

Comey, David Dinsmore. *Nuclear Power Plant Reliability: The 1973-74 Record.* BP 1-7507, Business and Professional People for the Public Interest.

Day, M. C. "Nuclear Energy: A Second Round of Questions." *Bulletin of the Atomic Scientists* (December 1975).

Des Moines *Register*, October 7, 1974.

Edison Electric Institute. *Report on Equipment Availability for the Ten-Year Period 1964-1973.* EEI Publication No. 74-57.

Electronics, "Solar Cell Offers 21% Efficiency with GaAs," May 29, 1975.

Forbes, "It Worked for the Arabs . . . ," January 15, 1975.

General Accounting Office. *Report to the Congress: Cost and Schedule Estimates for the Nation's First Liquid Metal Fast Breeder Reactor Demonstration Power Plant*, May 22, 1975.

General Advisory Committee to the U.S. Atomic Energy Commission, 130th Meeting, November 7, 1974, Uranium Supplies, Costs and AEC Resource Evaluation Program, Division of Production and Materials Management, U.S.AEC.

Huntington, Morgan. *The United States Nuclear Power Industry, a Summary of the Energy Yield and Fuel Demand.* A private memorandum, dated April 20, 1975.

INFO news release. *Atomic Industrial Forum* (March 1975).

Investor Responsibility Research Center, Inc. *The Nuclear Power Alternative.* Special Report 1975-A, January 1975.

Iowa, State of. Iowa State Commerce Commission, *In the Matter of Proposed Construction of Major Utility Plant, Statement of Intent.* Docket No. RES 75-1, August 19, 1975.

Joslin, Charles. "Nuclear Genie." *Barrons*, September 23, 1974.

———. "Uranium Enrichment," *Barrons*, July 7, 1975.

Kalhammer, F. R. and P. S. Zygielbaum. *Potential for Large-Scale Energy Storage in Electric Utility Systems.* ASME publication No. 74-WA 1 Ener-9, presented at the Winter Annual Meeting of the American Society of Mechanical Engineers, New York, November 17-22, 1974.

Klump, Dave. Project Manager, Solid Waste Utilization Project, Union Electric Company. Interview, January 1975.

Levitt, Theodore. *Innovation in Marketing: New Perspectives for Profit and Growth.* New York: Mc-Graw Hill Book Company, 1962.

"Low Marks for AEC's Breeder Reactor Study." *Science*, May 24, 1974.

Mitre Corporation, *Systems Analysis of Solar Energy Programs.* MTR-6513, report under contract to National Science Foundation, December 1973.

National Coal Association. *Bituminous Coal Facts, 1972.*

Olds, F. C. "Power Plant Capital Costs Going out of Sight." *Power Engineering*, August 1974.

Patterson, John A., Chief, Supply Evaluation Branch, Division of Production and Materials Management, ERDA. *U.S. Uranium Situation*. Speech given at the Atomic Industrial Forum Fuel Cycle Conference '75, March 20, 1975.

———. Speech given to the 17th Minerals Symposium of the American Institute of Mining, Metallurgical, and Petroleum Engineers at Caspar, Wyoming, May 11, 1974.

Price, John. *Dynamic Energy Analysis and Nuclear Power*. London: Friends of the Earth Ltd. for Earth Resources Research, Ltd., December 18, 1974.

Science, "Complications Indicated for the Breeder," Volume 185, August 30, 1974.

Temple, Jerry. Solid Waste System Superintendent, Ames, Iowa Power Plant. Interview, January 1975.

U.S. Atomic Energy Commission (AEC). *AEC Gaseous Diffusion Plant Operations*. ORO-684, January 1972.

AEC. *Environmental Survey of the Uranium Fuel Cycle*. WASH 1248, April 1974.

AEC. *Nuclear Fuel Resources and Requirements*. WASH 1243, April 1973.

AEC. *The Nuclear Industry*. WASH 1174: 14, 1974.

AEC. *Nuclear Fuel Supply*. WASH 1242, May 1973.

AEC. *Nuclear Power Growth 1974-2000*, WASH 1139 (74), February 1974.

AEC. *Statistical Data of the Uranium Industry*. GJO 650-100(74), January 1, 1974.

U.S. Environmental Protection Agency (EPA). *Energy Recovery from Waste*. EPA-SW-36d., 1973.

EPA. *Evaluation of Sulfur Dioxide Emission Control Options for Iowa Power Boilers*. EPA report #650 12-74-127, December 1974.

EPA., Office of Solid Waste Management Programs. *Energy Conservation Through Improved Solid Waste Management*. EPA-SWA-125, 1974.

EPA., Office of Solid Waste Management Programs. *Markets and Technology for Recovering Energy from Solid Waste*. EPA/530/SE-130, 1974.

EPA. *Report to Congress on Control of Sulfur Oxides*. EPA-450 11-75-110, February 1975.

U. S. Energy Research and Development Administration (ERDA), Office of the Assistant Administrator for Planning and Analysis. *Total Energy, Electric Energy, and Nuclear Power Projections*. February 1975.

ERDA. Uranium Enrichment Conference, Oak Ridge, Tennessee, February 13 and 14, 1975, CONF-750209.

ERDA, Office of Planning and Analysis, *Forecast of Nuclear Capacity, Separative Work, Uranium, and Related Quantities*, February, 1975.

U.S. Federal Energy Administration. *Project Independence Blueprint, Final Task Force Report: Coal*. November 1974.

U.S. Federal Energy Resources Council. *Uranium Reserves, Resources, and Production*, June 15, 1976.

U.S. Federal Power Commission. *Steam-Electric Plant Construction Cost and Annual Expenses*. Annual Supplements Nos. 14-25, (1961-72).

U.S. House of Representatives. Hearings Before the Subcommittee on Energy of the Committee on Science and Astronautics, June 6th and 11th, 1974.

Wall Street Journal, July 31, 1975.

———. June 7, 1976.

———. June 28, 1976.

———. July 15, 1976.

SAUNDERS MILLER is an investment banker specializing in mergers and acquisitions with the firm of Dain, Kalman & Quail, Inc. (member New York Stock Exchange) in Minneapolis. His career has included the analysis of hundreds of potential corporate investments and ventures.

He has previously been a financial consultant; has been Vice-President, Corporate Development of a major real estate developer, and Assistant to the Vice-President, Corporate Development of a Fortune 500 company. His experience has encompassed long-range strategic planning as well as specialization in the evaluation of companies as potential acquisitions. Among his activities in these positions, Mr. Miller has analyzed a broad spectrum of industries taking into account risk/reward ratios; performed in-depth financial analysis of companies together with assessing corporate operational capabilities, and performed market research. (The general concept of risk analysis used in this book is similar to the approach which Mr. Miller has used when representing buyers of companies.)

He has developed numerous profit and loss and cash flow forecasts for manufacturing companies as well as for existing and proposed real estate projects. His experience also has entailed the development of a corporate capital expenditure decision making model which included risk evaluation; designing and programming a simulation forecasting model for evaluating wage plans; and initiating management information systems in the areas of inventory control, labor reporting, engineering record control, quality assurance and sales analysis. Mr. Miller has also been an industrial engineer improving labor productivity and plant efficiency.

Mr. Miller received his Master of Business Administration from the University of Southern California, and his A.B. from The College of William and Mary in Virginia.

CRAIG SEVERANCE is a student at Iowa State University studying Public Administration and Business Administration.

ALTERNATIVE ENERGY STRATEGIES: Constraints
and Opportunities

John Hagel, III

ENERGY, INFLATION, AND INTERNATIONAL RELATIONS:
Atlantic Institute Studies II

Curt Gasteyger
Louis Camu
Jack N. Behrman
Foreword by Pierre Uri

FINANCING THE GROWTH OF ELECTRIC UTILITIES

David L. Scott

MANAGING SOLID WASTES: Economics, Technology,
and Institutions

Haynes C. Goddard

THE DYNAMICS OF ELECTRICAL ENERGY SUPPLY AND
DEMAND: An Economic Analysis

R. K. Pachauri